CONSENSUS ON
OPERATING PRACTICES
FOR THE CONTROL OF FEEDWATER
AND BOILER WATER CHEMISTRY IN
INDUSTRIAL AND INSTITUTIONAL BOILERS

prepared by the
Water Technology Subcommittee of the
ASME RESEARCH AND
TECHNOLOGY COMMITTEE ON
WATER AND STEAM IN
THERMAL SYSTEMS

THE AMERICAN SOCIETY OF MECHANICAL ENGINEERS
Two Park Avenue, New York, NY 10016

Preface

The Water Technology Subcommittee of the ASME Research and Technology Committee on Water and Steam in Thermal Systems, under the leadership of Mr. Robert D. Bartholomew has revised the *Consensus on Operating Practices for the Control of Feedwater Boiler Water Chemistry in Modern Industrial Boilers*, first published in 1979 with prior revisions published in 1994 and 1998. The task group consisted of a cross section of manufacturers, operators, chemical treatment contractors and consultants involved in the fabrication and operation of industrial and institutional boilers. Members of this group are listed in the acknowledgments.

This current document is an expansion and revision of the original, with reordered and modified texts where considered necessary. While significant revisions have been incorporated, it is recognized that there are areas of operating practice not addressed herein. Additional information is available from the references. It is the plan of the ASME Research Committee to continue to review this information, and revise and reissue this document as necessary to comply with advances in boiler design and water conditioning technology.

ACKNOWLEDGEMENTS

The current version of this document was revised, updated, and expanded by the following members of the Water Technology Subcommittee.

Robert Bartholomew, Chair

Aizaz Ahmed

Anton Banweg

Edward Beardwood

James Bellows

Wayne Bernahl

Deborah Bloom

E. Purusha Bonnin-Nartker

Kevin Boudreaux

Kirk Buecher

Irv Cotton

Bob Cunningham

David Daniels

Douglas DeWitt-Dick

Jim C. Dromgoole

Virginia Durham

Keith Fruzzetti

Frank Gabrielli

Robert Holloway

Julius Isaac

John Jevec

Jerry H. Jones

Eric Kangas

Jack Kelly

Roger W. Light

Lee Machemer

Chuck Marks

Tom Madersky

Peter Midgley

Elmer Mitchell

William Moore

Vickie Olson

George Patrick

Tom Przybylski

Jim Robinson

Anthony Rossi

Devin Ruth

Colleen Scholl

Joe Schroeder

K. Anthony Selby

Kevin Shields

Kumar Sinha

Merrill Willett

As most of the original Consensus document was retained, it is appropriate to present the names of the Feedwater Quality Task Group, which prepared prior versions of the Consensus on Operating Practices for the Control of Feedwater Boiler Water Chemistry in Modern Industrial Boilers

Robert T. Holloway, Chairman

Jesse S. Beecher

Wayne E. Bernahl

Deborah M. Bloom

Irvin J. Cotton

Robert J. Cunningham

Douglas B. DeWitt-Dick

S. B. Dilcer, Jr.

Arthur W. Fynsk

C. R. Hoefs

R. W. Lane

Jerome W. McQuie

D. E. Noll

Charles R. Peters

F. J. Pocock

James O. Robinson

Joseph J. Schuck

K. Anthony Selby

J. W. Siegmund

David E. Simon II

P. M. Thomasson

T. J. Tvedt, Jr.

J. F. Wilkes

CONTENTS

SECTION 1: INTRODUCTION

This document has been prepared by the Water Technology Subcommittee of the ASME Research and Technology Committee on Steam and Water in Thermal Systems as a consensus of proper current operating practices for the control of feedwater and boiler water chemistry in the operation of industrial and institutional, high duty, primary fuel fired boilers. These practices are aimed at minimizing corrosion, deposition, cleaning requirements, and unscheduled outages in the steam generators and associated condensate, feedwater and steam systems for boilers, and steam system components which are currently available. This publication is an expansion and revision of the operating practice consensus documents previously issued by the Committee [1-3]. The tabulated values herein update and replace the ones previously published. Titles have been edited and clarified. The text has been reordered and modified where necessary. **THE TEXT IS OF PRIME IMPORTANCE AND SHOULD BE CONSIDERED FULLY BEFORE USING THE TABULATED VALUES.** One Appendix has been added to provide additional guidance.

As designs and operation of equipment change, the Research and Technology Committee plans to review, revise, and reissue this document as necessary to include advances in boiler design or water treatment technology. This document is a consensus to use as a starting point to set up treatment programs, but standards can be set more or less stringently based on the recommendations of experienced water treatment personnel. This document should not be considered the ultimate authority.

The following is a list of other consensus documents developed by the Research Committee, plus two water-treatment-related performance test code (PTC) documents available from the ASME. The user is encouraged to obtain and review these other documents to ensure that the most appropriate document is used for a particular application. For example, a separate consensus has been developed to cover gas turbine heat recovery steam generators.

- Consensus on Operating Practices for Control of Water and Steam Chemistry in Combined Cycle and Cogeneration Power Plants, ASME Order No. 859988, 2012 [4].

- Consensus on Pre-Commissioning Stages for Cogeneration and Combined Cycle Power Plants, ASME Order No. 802485, 2017 [5].

- A Practical Guide to Avoiding Steam Purity Problems in the Industrial Plant, ASME CRTD Vol. 35, 1995 [6].

- Consensus on Operating Practices for the Sampling and Monitoring of Feedwater and Boiler Water Chemistry in Modern Industrial Boilers, ASME CRTD Vol. 81, 2006 [7].

- Consensus for the Lay-up of Boilers, Turbines, Turbine Condensers and Auxiliary Equipment, ASME CRTD Vol. 66, 2002 [8].

- Consensus on Best Tube Sampling Practices for Boilers and Nonnuclear Steam Generators, ASME CRTD Vol. 103, 2014 [9].

- Deaerators, ANSI/ASME PTC 12.3-1997 [10] (Performance Test Code), ASME Order No. D02397.

- High Purity Water Treatment Systems, PTC 31-2011 [11] (Performance Test Code), ASME Order No. C01611.

SECTION 2: BOILERS COVERED BY THIS STANDARD

The tables provided herein, and the seven classes of boilers covered in this document are provided in the following lists. The tables and accompanying notes for each should be considered together when using the data presented.

- Table 1: Industrial watertube, primary fuel fired, drum type with superheaters and turbine drives and/or process restrictions on steam purity. Feedwater with a reduced content of total dissolved solids (TDS) is typically required to meet these limits. Solids reduction can include maximizing condensate recovery and a higher degree of makeup treatment than softening..This boiler class excludes heat recovery system generators installed in gas turbine exhaust systems [See Reference 4].

- Table 2: Industrial watertube, primary fuel fired, drum type without superheaters and/or process restrictions on steam purity

- Table 3: Industrial firetube, primary fuel fired

- Table 4A: Industrial typically coil type, watertube, primary fuel fired steam generators with internal boiler recirculation pump

- Table 4B: Industrial typically coil type, watertube, primary fuel fired steam generator using feedwater pump for forced circulation through boiler

- Table 5: Marine propulsion, watertube, oil fired, drum type

- Table 6: Electrode type, high voltage, recirculating jet type

- Table 7: Process waste heat boilers providing steam to condensing steam turbines and turbine drives.

Each of these boiler types is defined in the Glossary. For the preceding tables, values are stated for typical, high duty design boilers with locally high heat fluxes not exceeding 150,000 Btu/hr/ft^2 (473.2 kW/m^2) and adequate water circulation. For older design units with lower heat flux, larger furnaces, larger steam drums, and very good water circulation, it may be sufficient to use targets given for a lower pressure range, especially where experience has indicated the success of such practices. In some cases, more restrictive targets may be required. The information also applies to steam generators in continuous or relatively steady-state operation. Special operating conditions such as startup, shutdown, rapidly fluctuating loads, or initial operation of new boilers may require greater water chemistry restrictions.

Tables 1 or 2 also may be sufficient for some rapid start boilers described by the criteria in the tables. However, natural circulation, direct-fired, rapid-start water tube boilers designed with low fluid velocities and horizontal tubes could pose unique challenges to those attempting to develop suitable limits. As indicated by (b) and (c) in the following quote, Klein et al., reported that "Major design factors which contribute to corrosion are: ... (b) horizontal or near horizontal tubes heated from above, and (c) horizontal or near horizontal tubes heated from below."[69] While these conclusions were based on surveys of high pressure boilers, laboratory studies of horizontal boiler tubes operating at 150 psig (1.03 MPa), 225 psig (1.55 MPa) and 250 psig (1.72 MPa) experienced corrosion in the tops of the horizontal tubes at a wide range of boiler water pH values (9.6-12.0) when inlet fluid velocities decreased below a critical value (e.g., <2-3 ft/s or <0.6-0.9 m/s in the examples shown) [70]. Tube temperatures and corrosion increased with proximity to the top of the tube and at locations of greater steam qualities [70]. These are only examples and should not be used to evaluate any specific boiler design. There are more parameters involved in boiler design such as tube diameter, heat flux, mass flux which can affect boiler circulation.

Operating practices are not given for the following classes of steam generators. Design, operation, and treatment of these types of equipment is too varied to permit the inclusion of consensus values:

- Combustion turbine heat recovery steam generators (HRSG)
- Mobile locomotive boilers
- Boilers of nonferrous materials (e.g., copper,aluminum, or other unusual materials)
- Immersion type, electric boilers, and low voltage electrode type boilers
- Heating boilers of special construction
- Firetube boilers with external superheaters
- Hot water boilers
- Oil field steam flood boilers and once through steam generators

- Solar boilers

- Secondary quench exchangers

Consult the referenced document for information on chemistry guidelines for combustion turbine heat recovery steam generators [4].

SECTION 3: OBJECTIVES OF WATER TREATMENT

The absence of adequate external and internal treatment can lead to operational upsets or unscheduled outages and is ill-advised from the point of view of safety, economy, and reliability. Where a choice is available, the reduction or removal of objectionable constituents by pretreatment external to the boiler is always preferable to, and more reliable than, management of these constituents within the boiler by *internal chemical treatment*, which involves boiler blowdown and chemical feed to the boiler system. The primary objectives of water treatment are as follows:

- Control steam purity
- Control boiler corrosion
- Control boiler deposition

Boiler blowdown helps to control steam purity, boiler corrosion and boiler deposition. Typically, a high continuous blowdown rate helps to achieve these three water treatment objectives. However, reducing blowdown is also a goal to minimize use of water, energy, and treatment chemicals. Section 3.4 provides additional guidance on blowdown.

3.1 CONTROL OF STEAM PURITY

The steam purity required for any given system is dictated by the intended use of the steam, and the type and concentration of steam contaminants permissible for this use. Consequently, the feedwater and boiler water contaminant concentrations must be limited so that the steam being produced is of satisfactory purity for its intended use. The steam purity targets suggested in Tables 1 through 5 and 7 are chosen to reflect the minimum requirements of an industrial steam use for each category of boiler operation and steam use.More stringent steam purity may be desired or required for some facilities within each category. Additional valuable information on these topics is also available in the referenced literature [12-34].

For Tables 1-5, the suggested steam purity targets are presented in terms of parts per million (ppm) and milligrams per liter (mg/L) of total dissolved solids (TDS). In practice, these targets are usually monitored by measuring the steam sodium and multiplying the steam sodium by the ratio of boiler water TDS to boiler water sodium to assess the amount of mechanical carryover and steam TDS concentration. Guidance on steam sampling system design is provided in ASTM D1066, Standard Practice for Sampling Steam [66]. See Reference 29 for information on carryover testing.

Table 7 provides a steam sodium limit of 10 micrograms per liter (µg/L) or parts per billion (ppb) in lieu of a TDS limit. This is primarily based on the steam sodium requirement for high pressure steam turbines. However, a ≤5 ppb steam sodium limit was the Central Electric Generating Board (CEGB) saturated steam sodium limit for boilers generally operated on low level caustic treatment and having austenitic alloys in the steam path (p. 1393 of Ref. [19]). More stringent steam sodium purity limits can be required for power industry boilers and combined cycle units [4].

In Table 1, steam TDS limits are <0.3 ppm to limit deposit accumulation in steel superheaters [6]. One boiler manufacturer recommends limiting steam TDS to <0.1 ppm to avoid deposits and corrosion of superheaters [p. 42-12 of Reference 24]. For boilers operating over 751 psig (5.18 MPa), a lower steam TDS limit (≤0.1 ppm) was recommended to avoid harmful deposits in superheaters. For boiler systems with steam turbines, "A Practical Guide to Avoiding Steam Purity Problems in the Industrial Plant"suggests limiting steam TDS to ≤0.06 ppm, steam sodium to ≤20 ppb, and steam silica to ≤20 ppb [6]. Superheated steam sodium targets of <10 ppb as Na (approximate TDS of <0.03 ppm) have also been recommended for industrial steam turbines.

Boiler manufacturers may be capable of fabricating boilers with lower carryover rates. It is understood that part of this better performance is due to the factor of safety used in the design by the boiler manufacturer (e.g., factor of 2 by one supplier) to ensure meeting steam purity guarantees. Operators of boilers can reduce carryover rates by limiting the steam generation rate. While the American Boiler Manufacturers Association (ABMA) presents fractional carryover values, their values are rounded to only one significant figure. Table 1 of the Consensus is based on carryover rates calculated directly from ABMA values for TDS in boiler water and steam. Boilers consistently operating with lower boiler water carryover rates may maintain proportionally higher boiler water conductivities. Similarly, boilers experiencing higher mechanical carryover rates would require proportionally lower boiler water conductivities. Table 1 includes steam sodium values which are estimated to roughly correspond to the TDS values shown. It is suggested that you measure the sodium and TDS in the boiler water

7

to calculate the sodium to TDS ratio for your system. As noted above, sodium is important due to the potential for caustic induced stress corrosion cracking on a steam turbine rotor. Chloride and sulfate in steam can result in stress corrosion cracking of stainless-steel alloys in downstream equipment, particularly equipment with high residual stresses (e.g., bellows-type expansion joints). The actual steam purity limits needed to avoid deposits or damage are specific to and set by the turbine manufacturer.

As indicated in the preceding paragraphs, more stringent targets than those in Tables 1-5 and 7 may be required to avoid turbine damage or other equipment damage. In cases where steam of greater purity is required than that obtained when operating within the targets suggested by the appropriate table, it is advisable to adjust the feedwater and boiler water chemistry suggestions as needed to achieve the desired steam purity. The actual permissible values for boiler water alkalinity, specific conductance, and silica may require adjustment to ensure compliance with steam purity targets.

Using a superheated (SH) steam sodium concentration limit (ppm Na_{SH}) to set the saturated steam sodium concentration (ppm Na_{SAT}), saturated steam TDS (ppm TDS_{SAT}) boiler water sodium (ppm Na_{BW}), boiler water TDS (ppm TDS_{BW}), and boiler water specific conductivity limit (SC_{BW}) involves the following steps. When spray attemperation is used, one must correct for sodium introduced by the attemperation water (ppm Na_{AT}) to determine the sodium concentration required in the saturated steam (ppm Na_{SAT}). In the following equations AT is the mass flow of attemperation water divided by the mass flow of superheated steam. All concentrations are in ppm.

$$Na_{SAT} \text{ ppm} = (Na_{SH} \text{ ppm} - AT \times Na_{AT} \text{ ppm})/(1-AT)$$

Then, one can calculate the TDS in steam and boiler water using the TDS/sodium ratio (RATIO, e.g., 3) for boiler water and the maximum mechanical carryover (MMC = % mechanical carryover/100%).

$$TDS_{SAT} \text{ ppm} = Na_{SAT} \text{ ppm} \times RATIO$$

$$TDS_{BW} \text{ ppm} = TDS_{SAT} \text{ ppm} / MMC$$

The corresponding boiler water specific conductivity limit can be calculated from the TDS/Conductivity Factor for the boiler water.

$$SC_{BW} \, \mu S/cm = TDS \text{ ppm} / (TDS/\text{Conductivity Factor ppm}/\mu S/cm)$$

This TDS/Conductivity Factor is assumed to be 0.65 ppm/μS/cm for Table 1. For example, the steam sodium concentration predicted for a 601-750 psig boiler by Table 1 is 100 ppb as Na (0.1 ppm Na). Assuming that no attemperation water is applied, the saturated steam purity needed is the same as the superheated steam sodium required. If a steam sodium concentration of <20 ppb as Na (0.020 ppm as Na)is needed for this pressure range, the 0.05% carryover rate would indicate a maximum boiler water sodium limit of 40 ppm as Na (0.020 ppm / 0.050%=40). Using a nominal sodium to TDS ratio of 3 ppm TDS/ppm sodium, this would correspond to 120 ppm of TDS. Using a conservative Specific Conductivity to TDS factor of 0.65 ppm/μS/cm, 120 ppm of TDS corresponds to 184.6 μS/cm (120 ppm / 0.65 ppm/μS/cm).

For this example, the maximum boiler water specific conductivity may need to be reduced from <920 μS/cm to <180 μS/cm (which is about 1/5 of value in Table 1) to ensure a sodium concentration of <40 ppm as Na in the boiler water and <0.020 ppm (<20 ppb) as Na in the steam. More stringent boiler water TDS limitations require higher purity feedwater and makeup water and/or a higher blowdown rate.

Tables 1, 4 and 7 also include steam silica limits. Total steam silica is the result of mechanical and volatile carryover of silica from the boiler water plus silica introduced with attemperating water. Boiler water silica targets were updated to be consistent with the silica control curves currently in use. See Appendix A to refine the silica target for a particular boiler.

Where direct spray water is added to steam for attemperation, the purity of the spray water must be consistent with downstream uses of the steam. Ideally, the purity of the spray water would satisfy the steam purity requirements. However, where this is not possible, attemperation water impurities should be controlled at concentrations that do not prevent the steam from meeting the steam purity targets. Nonvolatile treatment chemicals (e.g., caustic, sulfite, phosphate, polymer, and others) must not be applied to water that will be used for spray attemperation.

Sections 4.5 and 4.12 briefly discuss steam silica and sodium testing and their limits. For more information on these topics as well as the sampling and monitoring of steam purity, the reader is referred to ASME CRTD-Vol. 35, "A Practical Guide to Avoiding Steam Purity Problems in the Industrial Plant" and ASME CRTD-Vol. 81, "Consensus on Operating Practices for the Sampling and Monitoring of Feedwater and Boiler Water Chemistry in Modern Industrial Boilers" [6,7].

3.2 CONTROL OF CORROSION

Dillon, Desch and Lai provide a sufficient overview of corrosion mechanisms for industrial boilers and to a lesser extent condensate, feedwater, and steam system components [35]. Corrosion of feedwater system components and economizers is primarily inhibited by controlling feedwater pH and dissolved oxygen concentrations. The minimum recommended feedwater pH has been raised from 8.3 (ASME CRTD Vol. 34, 1998) to 8.8 in Tables 1, 2 and 7 to reduce steel corrosion. Flow Accelerated Corrosion (FAC) has caused a number of failures, including some with fatalities, of feedwater system components for boilers operating with demineralized quality makeup [36]. FAC can also occur in systems without demineralized makeup if they have a very large proportion (e.g., 90%) of high purity condensate return flow resulting in feedwater conductivities comparable to systems with demineralized makeup. Corrosion of condensate and feedwater system components produces corrosion products that can deposit on boiler tube surfaces and contribute to under deposit corrosion. In addition to the limits shown in the tables, more restrictive maximum and minimum normal operating ranges may need to be set for boiler water pH, P-alkalinity, and/or hydroxide alkalinity to control boiler corrosion. Due to the numerous variations in design, operation, makeup, process condensate, and treatment programs these limits should be tailored for each installation by someone sufficiently experienced in the art of boiler water treatment. The following references provide guidance on this topic [31, 36-51]. Improper or inadequate layup has often caused significant damage. For boilers that are out of service frequently or for extended periods, layup practices are required to control corrosion [8].

3.3 CONTROL OF BOILER DEPOSITION

Excessive deposition of corrosion products or insoluble contaminant species, such as hardness salts, on boiler internal surfaces, particularly boiling heat transfer surfaces, can result in the failure of boiler tubes due to overheating or under-deposit corrosion mechanisms. The potential for damage due to deposits increases with increasing operating pressure. Higher pressure boilers are much more susceptible to under-deposit corrosion. Deposits also severely limit heat transfer which can increase fuel costs or cause problems with components that will be exposed to or reach higher temperatures,

10

even when the deposits do not result in boiler tube failures in the near term. The targets suggested are in recognition of the great difficulty of effectively managing high concentrations of depositing contaminant species with internal treatment and deposit control agents alone, especially at elevated pressures.

With modern pretreatment technology as well as effective chemical treatment to control steam-condensate and feedwater corrosion, the targets for iron, copper, hardness, suspended matter, and other deposit forming materials are achievable under routine operating conditions. In industrial operations the potential for contamination and pretreatment failures and upsets is real, and such incidents may occur periodically [52]. As discussed further in the following paragraphs, it is extremely important for the boiler operator to remain alert and react properly to such upsets to minimize the potential for accelerated and damaging rates of deposition and/or corrosion. In addition, automated and real time monitoring of feedwater and boiler water quality plus key indications of contamination can be a valuable aid to the operator in sensing and properly reacting to upset/contamination events.

Inspection of fuel-fired boilers should include removal of boiler tube samples from the high heat transfer surfaces of the boiler to determine specific deposit weight or mass per unit of area on these surfaces, as well as the elemental deposit composition. Consult ASME CRTD-Vol. 103, "Consensus on Best Tube Sampling Practices for Boilers and Nonnuclear Steam Generators" [9].

When tube sample removal is inappropriate, certain nondestructive inspection techniques can provide useful information on boiler cleanliness. Video records of internal inspections provide a valuable history of internal boiler condition, and a basis for comparing current conditions to past inspections. Modern videoscope and boroscope recordings of key heat transfer and steam generating tube surfaces are particularly valuable for future reference and comparison. Nondestructive methods are inherently qualitative and potentially subject to misinterpretation. Analyses of tube samples provides a more reliable and quantitative indication of boiler tube cleanliness.

The suggested feedwater purity targets were established assuming an annual inspection frequency. More stringent limits may be required if considering longer inspection intervals. Shorter inspection intervals may be appropriate if contamination has been greater than normal.

3.4 CONTROL OF BOILER BLOWDOWN AND CHEMICAL FEED

Blowdown should be controlled to achieve stable boiler water treatment chemical residuals and baseline contaminant concentrations. The blowdown rate should be adjusted as required to control specific conductivity, silica, and alkalinity within the normal operating control targets for the boiler water. There are two commonly acceptable means of blowdown control:

- Boiler water specific conductivity control range. This is suggested for boilers with automated blowdown and chemical feed systems in which the boiler water treatment chemicals may be fed at a constant dosing rate based on the feedwater flow. The blowdown flow rate is regulated to automatically maintain a specific conductivity range in the boiler water. If there is a feedwater chemistry upset such as hardness, the chemical feed setting may be temporarily adjusted until the source of contamination is found and eliminated. The blowdown rate may also have to be increased temporarily to purge the contaminant from the boiler. Recommended silica and alkalinity targets corresponding to the operating pressure must be considered when establishing the blowdown specific conductivity control range. Some boiler systems may require blowdown control for silica or alkalinity rather than specific conductivity to maintain all contaminant concentrations within the recommended range.

- Fixed flow blowdown rate. This approach is very common in high pressure boilers with a constant steaming rate but has also been used in boilers with variable steam flow and minimal incidental contamination. The boiler chemical feed rate may be either automatically or manually controlled but is typically maintained in proportion to the feedwater flow rate. In such facilities, the boiler water chemistry is mainly dictated by the treatment chemicals, so it is important to keep the chemical feed rate and blowdown rate balanced. A noticeable change in boiler water chemistry typically indicates a contamination event in the feedwater. Changes in blowdown and chemical feed are implemented temporarily in response to the upset to stabilize chemistry until the contamination is eliminated.

Blowdown is increased to reduce the boiler water specific conductivity and cycles of concentration. Higher blowdown rates increase water and energy losses. Lower blowdown rates result in higher holding times, potentially increasing the risk of depositing particulates and other trace contaminants.

Specific conductivity is often used in estimating boiler cycles of concentration. This approach may be valid, but errors can be induced due to changes in alkalinity between the feedwater and boiler water and the effect of treatment chemicals applied. Where a significant portion of the feedwater specific conductivity is due to ammonia or amines, specific conductivity is not a reliable parameter for calculating cycles of concentration. For softened water supplies, monitoring naturally occurring chloride concentration in the boiler water and feedwater can be a good approach to calculate cycles of concentration. Inert, nonvolatile tracer chemicals are often added as part of the boiler water chemical treatment and can be used to determine cycles of concentration.

Prudent cycle limitations are dependent on feedwater purity and steam purity requirements. The feedwater purity is primarily dependent on the purity and proportion of makeup, and the proportion and purity of the condensate. However, industrial boilers with high purity makeup treatment systems, and satisfactory condensate contaminant concentrations may maintain a high feedwater purity, but a minimum of 1% blowdown, or 100 cycles of concentration is still recommended. This is to avoid excessive concentration of trace contaminants and the possible formation of deposits in the boilers. Operation at a maximum of 50 cycles reduces retention time and helps avoid deposition in industrial boilers. Boilers with a significant proportion of softened makeup will often require much higher blowdown rates, e.g., 2-10% of feedwater flow rate.

SECTION 4: WATER CHEMISTRY PARAMETERS

The maintenance of specified feedwater and boiler water chemistry must be well regulated and documented by frequent analysis and record keeping. Normally, a combination of online analyzers and grab sample measurements is used to ensure proper chemistry control. Guidance on sample collection and conditioning is provided in "Consensus on Operating Practices for the Sampling and Monitoring of Feedwater and Boiler Water Chemistry in Modern Industrial Boilers" [7].

Conventional units are shown first followed by metric units, where appropriate. The metric units of measurement follow the guidelines originally set forth in ASTM Designation E 380 [53]. Where there were questions regarding unit conventions those presented in NIST Guide for the Use of the International System of Units (SI), NIST Special Publication 811 2008 Edition, 2008 were used [67]. Some of these units may be unfamiliar to the United States reader, but their equivalence to the more familiar English units is clearly indicated by the accompanying presentation of all values in both systems of measurement. Tests designed to determine concentration technically should have units of mass per unit volume, e.g., mg/L or µg/L. However, it is customary in North America to report concentrations in units of ppm (where 1 mg/L is approximately equal to 1 ppm) and ppb (where 1 µg/L is approximately equal to 1 ppb). Errors introduced from this approximation are negligible for cooled samples discussed in this guide. Citation of current standards does not preclude the use of procedures and technologies which may be generated in the future.

4.1 SPECIFIC CONDUCTIVITY AND TOTAL DISSOLVED SOLIDS (TDS)

Suggested values for boiler water total dissolved solids (TDS) as blowdown control are expressed as unadjusted specific conductivity in microsiemens/cm (µS/cm) at 25°C because current practice is to use a conductivity meter to indicate boiler water solids concentration. The value has often been expressed as ppm (mg/L) dissolved solids using an integral conversion factor in the measuring instrument or by multiplying the conductivity reading by an external conversion factor. If such conversion is necessary to comply with past practice, it can be obtained by multiplying the specific conductance by a factor established empirically by gravimetric analysis. For unadjusted specific conductance this factor is typically 0.5-0.7, whereas 0.75-0.8 is typical for neutralized specific conductance. The TDS values in the ABMA standards [21] were expressed as ppm (mg/L) actual solids and not as ppm (mg/L) of some arbitrarily selected salt such as sodium chloride. Therefore, in order to establish a TDS to conductivity relationship for any individual case, it is necessary to measure actual TDS by a gravimetric determination of the evaporated residue, including any water of hydration not liberated by normal evaporation at 103°C. A typical relationship using this technique is 0.65 and was used in Table 1 and in Section 3.1. However, the system specific value would need to be determined empirically and will change with variations in the composition of boiler water dissolved solids.

The boiler water specific conductivity targets shown for Table 2 were based on the original maximum ABMA limits for boiler water TDS divided by 0.65 ppm/μS/cm. Table 1 is also based on 0.65 ppm/μS/cm but presents lower limits because they are based on more restrictive steam purity requirements for superheaters, turbine drives, or process restrictions. All specific conductivity limits in Table 1 were rounded to the nearest 10 μS/cm, except for the bottom two pressure categories. The percent carryover rates shown in Table 1 are rounded to the nearest 0.001% and were calculated from the original ABMA limits for steam and boiler water TDS. Section 3.1 presents a method for tailoring the boiler water specific conductivity limits to achieve a specific steam TDS or sodium concentration.

For boilers with softened makeup, significant free hydroxide alkalinity in the boiler water, and operating at pressures up to 750 psi (5.147 MPa), the TDS/Conductivity Factor can be 0.50 ppm/μS/cm. For these boilers, the corresponding maximum specific conductivity could be 1.3 times the values shown in Table 1. For example, the specific conductivity limits for low pressure boilers operating with TDS/conductivity ratios of 0.50 ppm/μS/cm are as follows:

- 300 psi (2.07 MPa)   2100 μS/cm
- 450 psi (3.10 MPa)   1800 μS/cm
- 600 psi (4.14 MPa)   1500 μS/cm
- 750 psi (5.17 MPa)   1200 μS/cm

If these higher limits are used, acceptable carryover rates should be verified and the TDS/Conductivity Factor should be confirmed through laboratory testing.

As stated in the tables, the specific conductance values are without prior neutralization and are expressed as µS/cm. The practice of converting a sample to its neutral salt form before measuring conductivity, i.e., neutralized conductivity, in order to provide a uniform TDS to conductivity ratio is considered unnecessary in most cases because the alkalinity of the boiler water is normally relatively constant and the conductivity range for blowdown control is quite broad, especially at pressures below 900 psig (6.21 MPa). Excess neutralization of a low TDS, low conductivity water might result in a higher measured conductivity. This has also been noted in water with significant amounts of organic anions. In addition, when boilers are equipped with instrumental monitors or controllers for blowdown control, such instruments usually read directly in µS/cm of unadjusted specific conductivity. All targets should be based on direct reading of specific conductivity. Other units commonly used in industry include micromhos/cm (µmhos/cm), millisiemens per centimeter (mS/cm) and millisiemens per meter (mS/m). The conversion factors for conductivity are as follows 1 µS/cm = 1 µmhos/cm = 0.001 mS/cm = 0.1 mS/m = 0.0001 S/m.

4.2 pH

Feedwater pH

The suggestions for feedwater pH are based on values that will protect the pre-boiler system from corrosion and are consistent with the indicated pretreatment and internal boiler water treatment. For low pressure boilers that do not use feedwater for direct spray attemperation, nonvolatile alkali such as sodium hydroxide can be added to the feedwater to raise the boiler water pH. This will result in higher feedwater pH values that will reduce corrosion of steel surfaces between the point of injection and the boiler. Adding caustic to the feedwater may increase the amount of boiler blowdown required. Feedwater in installations with hot lime softeners may have pH values up to approximately 10.5.

For systems with high purity makeup water, the increase in pH to attain the tabular feedwater pH values is to be accomplished using volatile alkaline materials only (i.e., amines and/or ammonia). This limitation is consistent with the assumed use of high purity water and the corresponding assumption that the internal boiler water treatment will utilize some form of alkaline phosphate treatment. Obtain qualified guidance regarding treatment options and material selection for the steam/condensate system when neutralizing amine addition is not permitted.

Normal Operating Boiler Water pH Values

For industrial boilers, boiler water pH is usually controlled either by monitoring the pH directly or by monitoring and controlling the boiler water alkalinity. The relationship between alkalinity and pH can vary with the composition of makeup water supply and types of treatment chemicals. The boiler water pH shall be kept alkaline, and minimum and maximum normal operating limits should be established. The normal operating targets depend on the particular type of internal water treatment program selected.

When using feedwater that is primarily high purity condensate, demineralized makeup, and/or makeup treated by reverse osmosis, online boiler water pH monitoring should be used for control. If the makeup water is softened as defined in Section 5.1, boiler water alkalinities should be monitored and used for control.

An adequate explanation of corrosion and deposition control and factors considered for development of pH/alkalinity limits for all boiler water treatment programs and makeup/feedwater supplies would require a lengthy, complicated discussion not possible herein. Since that sort of treatise on water treatment is inconsistent with the goal of this consensus to provide a simple guide to operators, normal operating boiler pH limits are not presented in this document. However, a brief discussion of low upset pH conditions is included.

Minimum Boiler Water pH/Alkalinity for Firing

In general, if the boiler water pH cannot be maintained above 8.0 at 25°C the boiler should be removed from service to minimize the risk of low pH corrosion. There can be acceptable exceptions. For example, "TIP 0416-05 – Response to Contamination of High Purity Boiler Feedwater" is a guide specific to the pulp and paper industry that defines response to low pH values and allows a lower pH before immediately removing a boiler from service [54]. However, the minimum normal operating boiler water pH for most industrial boilers should be set well above 8.0 at 25°C.

P-alkalinity or OH-alkalinity is often used in lieu of pH to control the alkali content of the boiler water in low pressure boiler systems with high concentrations of alkalinity. Control is attained by analysis of grab samples. In such systems there usually is a large reserve of alkalinity which helps ensure chemistry readings remain stable during the periods between sequential sampling analysis sets. A positive P-alkalinity or OH-alkalinity by definition ensures a pH above 8.0-8.3 when measured at 25°C. Higher sample temperatures may result in significant testing error.

4.3 TOTAL ALKALINITY

The maximum boiler water alkalinity values given in Tables 1 through 4 and Table 6 are specified as total or methyl orange alkalinity, commonly abbreviated T or M alkalinity, and expressed in ppm (mg/L) $CaCO_3$ for all *boilers* operating below 900 psig (6.21 MPa). Samples are titrated with acid to pH 4.5 or the methyl orange indicator endpoint. The methyl orange endpoint can be difficult to see and has largely been replaced by other indicators such as methyl purple in practice and in the standard references. These indicators have a higher pH endpoint in the range of 4.9-5.2 but are considered to be sufficiently alike for most industrial boiler water testing. Total alkalinity correlates with pH, corrosion inhibition, and carryover tendency from foaming and was used in previous versions of this guideline [1-3, 21]. For boilers with conductivities in excess of 250 µS/cm, maximum boiler water total alkalinity limits in Table 1 were set at 20% of the estimated total dissolved solids to minimize foaming potential.

Section 3.1 and 4.1 present methods for adjusting boiler water specific conductivity limits based on carryover rates, steam purity requirements, TDS to sodium ratios and TDS to conductivity ratios. When the boiler water TDS limits are reduced, the total alkalinity limits should also be reduced. If the boiler water specific conductivity is >250 µS/cm, the maximum alkalinity limit should be reduced according to the following formula.

Maximum boiler water total alkalinity ppm as $CaCO_3$ = 0.2 x maximum boiler water TDS ppm.

4.4 HYDROXIDE ALKALINITY AND PHENOLPHTHALEIN ALKALINITY

The presence of phenolphthalein alkalinity (i.e., pH values in excess of 8.2 at 25°C) is generally required in all operating boilers to reduce the potential for acid corrosion in boilers, - although pH or

18

hydroxide alkalinity are commonly monitored and controlled instead of phenolphthalein alkalinity to demonstrate a sufficient minimum alkali level in the boiler water. Much higher levels of hydroxide alkalinity can be required in low pressure boilers to help fluidize sludge and inhibit deposition of silicates. However, underdeposit evaporative concentration of hydroxide at tube surfaces can lead to excessive amounts of hydroxide and caustic gouging (see glossary). While maximum hydroxide alkalinity limits may be shown, water circulation, heat flux, and deposition all influence the concentration reached at the tube surface. Much more stringent limits may be required to reduce the risk of caustic gouging in some boilers.

Tables 4A, 4B and 5 have included a maximum hydroxide alkalinity limit since the first published version of this consensus document. Hydroxide alkalinity values are given for the typically coil type watertube boilers, Table 4A and 4B, because the use of hydroxide to solubilize silica is more critical for boilers in this category.

Statements in some of the table-specific notes advise having specified minimum and maximum pH, P-alkalinity, or hydroxide alkalinity limits set by water treatment personnel for each boiler in order to ensure silica solubility and proper functioning of deposit control chemical treatments. Direct measurement by titration of boiler water hydroxide and/or phenolphthalein alkalinities is a commonly recommended practice for controlling pH in low pressure industrial boilers. Since hydroxide alkalinity and P alkalinity test procedures were not included in the ASME sampling document, a discussion of these procedures is provided herein. This is a simplified version of the discussion presented in Reference 45 [45].

Some proponents [42] control boiler water pH/alkalinity using Phenolphthalein alkalinity (P alkalinity) titrations. P alkalinity is measured by titrating a sample with acid to the P alkalinity endpoint, which occurs at a pH of about 8.0-8.3 with the indicator when measured at 25°C and at a pH of 8.3 (at 25°C) if a pH meter is used in lieu of an indicator according to Standard Methods [55].

For industrial boilers, barium chloride is the preferred method of titrating hydroxide alkalinity. Barium chloride is added to the sample and the sample is then titrated with acid to the P alkalinity end point. Barium precipitates all the carbonate and phosphate, thus leaving only hydroxide. The barium chloride method reportedly provides readings with an accuracy of 10% [39,56].

In North America, alkalinities are normally expressed as mg/L (ppm) $CaCO_3$. Elsewhere, they are expressed as equivalents per liter where 1 eq/L is equal to 50 mg/L $CaCO_3$. In the case of hydroxide alkalinity only, results are sometimes expressed as equivalents as NaOH, where 1 ppm NaOH = 1.25 ppm $CaCO_3$. However, in these consensus guidelines caustic or hydroxide alkalinity is expressed as ppm $CaCO_3$.

Maximum suggested hydroxide alkalinity targets have been provided for >900 psig boilers in Table 1. Direct measurement of pH is preferred in high pressure boilers operating with high purity feedwater and low solid treatment programs. See Section 4.2 for information on pH. The hydroxide alkalinity tests are not sufficiently accurate for hydroxide alkalinity concentrations below approximately 4 ppm as NaOH in boilers using phosphate treatment. Use of mathematical algorithms based on pH, phosphate, ammonia, and amine, if present, are preferred for estimating hydroxide alkalinity in these systems. Since amines are commonly used in industrial boilers in lieu of ammonia, the algorithms in "Consensus on Operating Practices for Control of Water and Steam Chemistry in Combined Cycle and Cogeneration Power Plants" [4] need to be modified. Guidance on modifying these formulas is provided by Reference 49.

4.5 SILICA

Maximum boiler water silica concentrations in the operating pressure ranges ≥901 psig (6.22 MPa) are generally selected so that total carryover will not exceed 20 ppb (20 µg/L) silica, as SiO_2, in steam according to the well-established silica volatility data of Coulter, Pirsh, and Wagner [13] and estimated mechanical carryover of silica. Appendix A provides a more accurate method for estimating boiler water silica limits to achieve a desired steam silica concentration for a particular boiler.

At lower operating pressure ranges (<900 psig or <6.22 MPa) in Tables 1-6, the boiler water silica values are selected to avoid internal deposition of complex silicates. This deposition might occur on heat transfer surfaces in fuel fired and waste heat boilers and on the spray nozzles in electrode boilers. The combination of silica with iron, aluminum, or hardness can concentrate within porous deposits and exceed the solubility of complex silicates. There is also a recommendation for fuel fired and waste heat boilers operating ≤900 psig (6.21 MPa) that pH or hydroxide alkalinity be maintained at a high enough level to ensure silica solubility. The boiler water silica concentration limits in Table 1 were reduced from the prior version of this table to ensure that one could maintain 2.5 ppm as $CaCO_3$ of hydroxide alkalinity per ppm of silica without exceeding the more stringent total alkalinity limits. Polymers tailored to inhibit silica deposition may enable higher silica levels without this level of hydroxide alkalinity. Table 8 presents the silica limits required to achieve ≤20 ppb (≤20 µg/L) silica in the saturated steam from these lower pressure boilers.

4.6 DISSOLVED OXYGEN

Dissolved oxygen concentrations are stated for feedwater samples drawn either before or after chemical oxygen scavenger addition as specified by each table. Where the dissolved oxygen concentration is stated as 7 ppb (µg/L) O_2 or less measured before chemical oxygen scavenger addition, it is assumed that a well-operated deaerator is in service. In all cases, the subsequent addition of a chemical oxygen scavenger to the deaerated water with adequate distribution and mixing is desirable. This should provide essentially zero dissolved oxygen in the feedwater at the economizer inlet or, in the absence of an economizer, at the feedwater inlet to the boiler.

For units with an all-volatile feedwater treatment program and an all-steel steam/water cycle/boiler system, flow accelerated corrosion (FAC) of susceptible components may be of significant concern at low dissolved oxygen concentrations with a volatile reducing agent or oxygen scavenger present [36]. Susceptible components may include feedwater lines, condensate lines, and heater drain lines. Guidance is available to assess and minimize the risk of FAC [36]. In such cases, the cycle may more closely match the chemistry of a combined cycle power plant and the reader is advised to consult and consider the AVT(O) treatment programs discussed in the ASME "Consensus on Operating Practices for Control of Water and Steam Chemistry in Combined Cycle and Cogeneration Power Plants" [4]. The Electric Power Research Institute (EPRI) also has excellent chemistry guidance documents for power plants [31].

Dissolved oxygen analyses consistent with the desired minimum level of detection should be made either by online monitors or through flowing grab sample analyses using appropriate standard methods [57-58]. On-line methods include electrochemical and optical technologies.

Care must be taken to minimize air ingress when testing for low concentrations of dissolved oxygen, especially for deaerated high purity water.

4.7 FEEDWATER CORROSION PRODUCTS (IRON AND COPPER)

The suggested targets for iron and copper in feedwater are set at low ranges, and these targets decrease progressively with increasing boiler pressure. Other metals such as zinc, nickel, aluminum, and manganese may also be present at elevated concentrations. While individual limits for these other metals are not provided, the sum of all corrosion products should not exceed the sum of the iron and copper limits. Minimizing the transport of these metals primarily serves to avoid deposits on boiler tube surfaces. Jet type electrode boilers are subject to erosion/corrosion of the electrodes by metallic precipitates in the boiler water that are recirculated at a high rate, and high concentrations of iron and copper may increase the possibility of ground fault arcing.

External pretreatment to minimize corrosion products is needed to achieve and maintain the recommended feedwater contaminant and boiler water chemistry targets cited in the following tables. As stated in several of the table specific notes, some internal treatment programs with chelants or polymers may permit reliable boiler operation with modestly higher concentrations of feedwater iron, copper, and hardness contaminants than those cited in the tables. These higher contaminant concentrations should be tolerated only after careful consideration by competent water treatment personnel. The acceptability of operating with the higher contaminant concentrations must be confirmed by routine internal boiler inspections and other acceptable deposition rate monitoring techniques [52-53, 59-60].

Colorimetric and/or spectrophotometric test methods are most frequently used. These rely on reagents that react with copper or iron to form colored compounds. The colored sample is then measured using instrumentation that passes light through the sample at wavelengths specific for the compound being measured. The amount of light absorbed by the sample determines the concentration of copper or iron in the sample. The primary disadvantage of colorimetric testing is the potential for interferences from other metals in the water. With the development of Atomic Adsorption (AA), Inductively Coupled Plasma (ICP), and various instrumentation enhancements, the Relative

Detection Limit (RDL) has improved and interferences have been minimized. See Table 4 of reference [7].

4.8 HARDNESS

Total hardness refers to the collective sum of calcium (Ca^{2+}), magnesium (Mg^{2+}), barium (Ba^{2+}), and strontium (Sr^{2+}) cations in water. Each of these cations can react with various compounds in a water to form insoluble deposits. Typically, only calcium and magnesium are present in natural waters in appreciable amounts. Hardness can precipitate and form deposits on heat transfer surfaces. The primary concerns with hardness deposits on the water side of steam generation tubes are the insulating effect of these deposits and the restrictions in boiler water flow; both of which can ultimately result in boiler tube failures.

The suggested targets for hardness in the feedwater are set so that the internal treatment program can effectively minimize deposits. Feedwater hardness below the recommended targets should be maintained where practical. Reduction in chemical usage and minimization of internal deposition will result in energy and operational cost savings, as well as improvements in long term life expectancy.

Softened water should comply with the feedwater hardness concentration limits shown in the Tables. Softened water is defined in Section 5.1.

With the development of higher degrees of pretreatment and more sensitive testing methods, the ability to measure very low hardness concentrations has improved substantially. Traditional colorimetric titration methods using EDTA and Eriochrome Black T have a practical lower detection limit of 0.1-0.25 mg/L (0.1-0.25 ppm) or 100-250 µg/L (100-250 ppb) depending on the operators' care and skill in sampling and testing, time delay between sampling and testing, analytical equipment, and the method used. Additional information is provided by Table 4 in ASME "Consensus on Operating Practices for the Sampling and Monitoring of Feedwater and Boiler Water Chemistry in Modern Industrial Boilers" [7]. Larger sample sizes up to 1000 mL may improve the relative detection limit (RDL), but the color change may be difficult to detect. Colorimetric spectrophotometric procedures are also available that can improve accuracy.

Instrumentation such as an atomic absorption spectrophotometer, inductively coupled plasma or ICP mass spectrometer, and ion chromatography (IC) have reduced the relative detection limit to <10 µg/L (0.01 ppm). See Table 4 in ASME "Consensus on Operating Practices for the Sampling and Monitoring of Feedwater and Boiler Water Chemistry in Modern Industrial Boilers" [7]. High purity water is defined as water having total hardness of ≤ 10 µg/L (≤ 10 ppb). For industrial plants operating with waters treated only by ion exchange softening, maintaining hardness concentrations <10 µg/L (≤ 10 ppb) in the feedwater is impractical or unachievable. If hardness cannot be measured or is higher than these concentrations, expert assistance from a treatment specialist should be obtained in order to understand the effect of hardness on the chemical treatment program being used.

4.9 NONTREATMENT, NONVOLATILE TOC

The types of organic matter that can be present in an industrial boiler feedwater are numerous and extremely varied. Organics may exist in the makeup water from natural sources, be added as part of the boiler water chemistry, or be present through contamination of makeup water or condensate. Therefore, it is impossible to define best practice conditions for all categories in all situations. It is necessary to know the plant well and to identify the potential for organic materials to enter the feedwater. If present constantly or periodically at significant concentrations, identify, define, and determine concentrations and durations for such organic materials. Ascertain means of detecting the presence and ways to prevent these from contaminating the feedwater. Not all nonvolatile total organic carbon (TOC) is a concern, however, so it's necessary to determine whether the material has a negative impact on the boiler, boiler water chemistry, or steam purity and formulate action plans for contamination events as needed.

The following types of organic materials are examples of contaminants that can contribute to carbon-based deposits in boilers.

- No 6 fuel oil. See Section 4.10
- Black liquor
- Sugar

Organic matter can also decompose to acidic compounds and cause low pH values in boiler water. For example, this has been known to occur from starch contamination of boiler water. Severe starch contamination of condensate has been detected with chemical oxygen demand (COD) tests, although that test is not specific for starch. Several factors can influence decomposition of organic carbon in steam/water cycles [61].

The tables list values for nonvolatile TOC. This analysis in not defined by any published standard method. It is intended to represent a reasonable approach to the determination of organic feedwater contaminants potentially damaging to the boiler. Nonvolatile TOC measurement is an unofficial modification of the TOC test [62] conducted on a sample after atmospheric boiling with the subsequent subtraction of a calculated carbon value equivalent to the carbon content of any nonvolatile organic treatment chemicals. For more information, consult ASME "Consensus on Operating Practices for the Sampling and Monitoring of Feedwater and Boiler Water Chemistry in Modern Industrial Boilers" [7].

Nonpurgeable organic carbon is not exactly the same as nonvolatile TOC, but it is a reasonable substitute analysis technique in some situations. Nontreatment, nonvolatile TOC includes oily matter, and the limits are not additive where both are shown.

4.10 OILY MATTER

Oily matter [63] is not restricted to petroleum oils. It includes all nonvolatile hydrocarbons, vegetable oils, animal fats, waxes, soaps, greases, and related matter, all of which are extractable in halogenated solvents at low pH values. This grouping, large as it is, excludes some potentially damaging organic feed water contaminants and includes some beneficial organic compounds that may be added intentionally as a feed water treatment. The historical limits for oily matter were retained for Table 5 because it is normally used in lieu of the nonvolatile TOC limits for these boilers.

Oily matter may result in foaming in the boiler drum and can be very serious. Heavy oils can also carbonize on tube surfaces and form an extremely adherent and difficult to remove deposit. Users of this document should review possible oily matter contamination sources that can get into their system. Do not overlook oil from pumps for makeup, condensate, feedwater, and chemicals. Organic contamination of the feedwater or boiler can be detected by methods for oily matter or nonvolatile TOC described in ASME "Consensus on Operating Practices for the Sampling and Monitoring of Feedwater and Boiler Water Chemistry in Modern Industrial Boilers" [7].

4.11 VOLATILE ORGANIC MATTER

Some volatile organic matter (VOM) may cause damage to turbines. Low concentrations of weak organic acids, such as acetic and formic, are frequently found in the steam/water systems of boilers and steam generators. Considerable debate continues as to whether these weak acids actually influence corrosion in portions of the system or are merely nuisance ions. Excessive concentrations of organic acids in boiler water, e.g., from process condensate, may contribute to boiler corrosion, but there was insufficient information available to include guidance for VOM in this document.

4.12 SODIUM AND POTASSIUM

Online steam sodium monitoring is used in many industrial steam systems to determine if the steam complies with turbine manufacturer's steam purity limits or other process constraints. The typical targets for industrial condensing steam turbines operating with superheated steam range from <3 µg/L to <20 µg/L (<3 ppb to <20 ppb) of sodium as Na. Some turbine suppliers specify that the sum of sodium (Na) and potassium (K) shall meet stipulated sodium limits. Unfortunately, online sodium analyzers do not detect potassium, and compliance with limits requires grab sample analyses to determine the actual ratio of boiler water potassium to sodium if significant potassium is present (e.g., ppm K>0.2 X ppm Na). Sodium analyzers are specific to sodium and are practically unresponsive to potassium.

Potassium salts can be much more corrosive than sodium salts to steel in boiler superheaters and generation bank tubes. Therefore, elevated concentrations of potassium in the boiler water should be avoided. See Section 3.1 for more information on steam purity.

4.13 CATION CONDUCTIVITY (CONDUCTIVITY AFTER CATION EXCHANGE)

While the tables do not provide any limits for cation conductivity, this terminology is defined in this document for industrial users in case it is listed in steam purity limits for new condensing turbine installations. Cation conductivity is often called acid conductivity or conductivity after cation exchange (CACE). It is the measurement of conductivity after all the cations in the sample have been exchanged for hydrogen ions. The standard instrument for measuring cation conductivity exchanges the cations by passing the sample through hydrogen form cation exchange resin. Another method based on continuous electrodeionization was recently introduced.

Cation conductivity provides an indication of anionic contamination present in the sample. Industrial steam systems can have large amounts of volatile anions such as carbon dioxide, which will result in high cation conductivity readings. If cation conductivity monitoring is required for assessment of compliance with the turbine steam purity limits, the option of monitoring degassed cation conductivity instead of cation conductivity should be considered. The two most common forms of degassing are nitrogen gas sparging and boiling the sample.

SECTION 5: TABLES OF SUGGESTED WATER CHEMISTRY TARGETS

Consensus water chemistry controls for the six types of steam generator systems are presented in Tables 1 through 7. The tabulated information is categorized according to operating pressure ranges because this is the prime factor that dictates the type of internal water chemistry employed, the normal cycles of feedwater concentration, the silica volatility, and the carryover tendency. The difference between steam and water densities decreases with increasing pressure and temperature; therefore, separating the steam/water phases completely in the boiler drum becomes increasingly difficult to achieve. Since the tendency to carryover is greater at higher operating pressures, it is necessary to maintain lower boiler water contaminant concentrations to meet the same steam purity target.

The tables are not categorized by the type of fuel used. Tables 1-6 apply to boilers fired with primary fuels such as oil, gas, or coal, but they have also been found to be applicable to some conventional waste heat boilers. Table 7 is for high duty process waste heat boilers. For heat recovery steam generators (HRSG) or waste heat boilers, consult the ASME "Consensus on Operating Practices for Control of Water and Steam Chemistry in Combined Cycle and Cogeneration Power Plants" [4]. As a caution, waste heat boilers are sometimes designed and operated so that waterside circulation is inefficient, areas of unavoidable deposit accumulation are numerous, and localized heat fluxes are abnormally high. In such instances, the waste heat units, regardless of their operating pressure, should be operated with high purity makeup water and chemistry should be consistent with the Table 1 values for boilers operating above 1000 psig (6.89 MPa). Boilers with significant circulation problems or other design deficiencies can eventually fail even with the best possible chemistry program. The point of maintaining excellent makeup purity and cycle chemistry control is for slowing the demise of equipment until the design deficiency can be corrected.

5.1 DEFINING MAKEUP PURITY FOR TABLES

Recommendation of specific types of makeup water pretreatment, condensate treatment, and internal chemical treatment is outside the scope of this document. However, the requirement for such treatments, in many cases, is clearly implied by the suggested values for feed water quality. Specific reference is made to such pretreatments as demineralization, evaporation, softening, dealkalization, softened and dealkalized, reverse osmosis, or continuous electrodeionization either where such treatments are common practice or where they describe the range of applicability of the control values in a certain table. While not specified or implied by the name, some high purity water or softened water supplies have a separate, makeup deaerator to degasify water before entering the condensate/feedwater cycle.

High Purity Water

For the purposes of the enclosed tables, high purity makeup water should have a specific conductivity ≤10 µS/cm. Many facilities will have water with conductivities well below this value and many boilers will require a much higher purity water specification to satisfy the requirements for feedwater, attemperation water, and steam purity. Hardness and iron shall be less than the feedwater targets in the appropriate table. Silica shall be sufficiently low to keep boiler water silica within the suggested target of the appropriate table at 1% blowdown. For boilers operating at ≥1500 psig and/or units operating on all-volatile treatment programs, specific conductivity should be ≤0.1 µS/cm and silica should be ≤0.02 mg/L silica, as SiO_2, in the effluent of the high purity makeup water system before storage. High purity water should also have no detectable hardness, ≤10 ppb as $CaCO_3$, per the discussion on hardness. Iron should be ≤0.01 mg/L as Fe. High purity water can be produced through a variety of treatment processes.

Softened Water

For the purposes of the enclosed tables, softened water is defined as a makeup water meeting the following requirements:

- Hardness and iron shall meet the feedwater target in the appropriate Table.

- If ion exchange softening is used, the specific conductivity of the softener effluent will be less than 1.2 times the influent specific conductivity provided the resin bed has been well rinsed following regeneration. Generally, the specific conductivity will be well above 10 µS/cm.

5.2 TYPES OF BOILER WATER CHEMICAL TREATMENT

Recommending a specific type of chemical treatment is beyond the scope of this guideline. The following provides a list of references for discussion of common feedwater and boiler water treatment programs.

- Coordinated phosphate-pH control [37]
- Congruent phosphate-pH control (a modification of coordinated phosphate treatment) [38]
- Equilibrium phosphate treatment [41] and other power industry phosphate treatments [4, 31, 48]
- Volatile treatment [31,47] (which is mainly used only in the power industry)
- Caustic treatment [31 – as it applies to power industry boilers]
- Other treatments for low pressure boilers (chelation, dispersion, crystal modification, threshold inhibition, and precipitation programs) [39, 40, 44, 51, 64, 65]

The one common aspect of these programs is the need to maintain boiler pH and alkalinity. See Section 4.2 and footnotes specific to each table.

5.3 USE OF THESE TABLES FOR CHEMISTRY CONTROL

For primary fuel fired boilers, it should be recognized that oil firing causes the greatest release of radiant heat in the furnace and this creates the most stringent limitations on deposit-forming materials entering the boiler with the feedwater. Coal firing releases less radiant heat while gaseous fuels release the least radiant heat.

Low pressure boilers frequently are operated with feedwater that is suitable for use in higher pressure boilers. In such cases the boiler water chemistry targets should be based on the pressure range that is most consistent with the feedwater chemistry. For example, if a boiler operated at 150 psig (1.03 MPa) uses feedwater of suitable quality for use in a 1001-1500 psig (6.9-10.34 MPa) boiler, then the boiler water targets and chemical treatment program should be based on the higher pressure guidelines. This practice is necessary to ensure proper blowdown and to avoid extremely high concentrations of trace contaminants and impurities leading to the formation of deposits in the boiler.

SUGGESTED WATER CHEMISTRY TARGETS INDUSTRIAL WATERTUBE WITH SUPERHEATER

TABLE 1   PRIMARY FUEL FIRED, DRUM TYPE

Makeup water percentage: Up to 100% of feedwater (1)

Conditions: Includes superheater. Process limitations may require more restrictive steam purity, boiler water and feedwater targets. Turbines usually require stringent steam purity limits (consult turbine manufacturer limits).

Drum Operating Pressure, (1, 10)	psig	0-300	301-450	451-600	601-750	751-900	901-1000	1001-1500	1501-2000
	(MPa)	(0-2.07)	(2.08-3.10)	(3.11-4.14)	(4.15-5.17)	(5.18-6.21)	(6.22-6.89)	(6.90-10.34)	(10.35-13.79)
Feedwater (2, 3, 4)									
Dissolved oxygen, ppm (mg/L) O_2 measured before chemical oxygen scavenger addition (2)		< 0.007	< 0.007	<0.007	<0.007	<0.007	< 0.007	<0.007	<0.007
Total iron, ppm (mg/L) as Fe (3)		≤0.10	≤0.05	≤0.03	≤0.025	≤0.02	≤0.02	≤0.01	≤0.01
Total copper, ppm (mg/L) as Cu (3)		≤0.05	≤0.025	≤0.02	≤0.02	≤0.015	≤0.01	≤0.01	≤0.01
Total hardness, ppm (mg/L) as $CaCO_3$ (3)		<0.5	≤0.3	≤0.2	≤0.2	≤0.1	≤0.05	≤0.01	≤0.01
pH @ 25°C		8.8-10.5	8.8-10.5	8.8-10.5	8.8-10.0	8.8-10.0	8.8-9.6	8.8-9.6	8.8-9.6
Chemicals for pre-boiler system protection		NS	NS	NS	NS	NS	VAM	VAM	VAM

Continue

Nonvolatile TOC (including oily matter), ppm (mg/L) as C (4)	<1	<1	<0.5	<0.5	<0.5	<0.2	<0.2	<0.2
Boiler Water (5, 6, 7, 8, 9, 10)								
Silica, ppm (mg/L) SiO_2 (10)	≤84	≤72	≤40	≤30	≤18	≤5.9	≤1.7	≤0.7
Hydroxide Alkalinity, ppm (mg/L) as $CaCO_3$ (5)	See Footnote 5					<10	<2	<2
Maximum Total Alkalinity, ppm (mg/L) as $CaCO_3$	<210(6)	<180(6)	<150(6)	<120(6)	NS(7)	NS(7)	NS(7)	NS(7)
Maximum Suggested Target Specific conductance, µS/cm @ 25°C without neutralization to comply with steam purity limits (8)	≤1600	≤1400	≤1150	≤920	≤230	≤190	≤150	≤80
Saturated Steam Purity Target (8, 9,10)								
TDS (maximum), ppm (mg/L)	<0.3	<0.3	<0.3	<0.3	0.1	0.1	0.1	0.1
Sodium Estimated from TDS, ppb (µg/L) as Na	<100	<100	<100	<100	<30	<30	<30	<30
Maximum Mechanical Carryover [Calculated from TDS in Ref 21], %	0.029	0.033	0.040	0.050	0.067	0.080	0.100	0.20
Superheated Steam Purity Target (8, 9,10,11)								
Sodium, ppb (µg/L) as Na	<20 ppb (consult turbine supplier or qualified consultant for appropriate values for the system)							

NS = not specified, VAM = volatile alkaline materials.

NOTES FOR TABLE 1

1. Most new and existing boilers can and should have feedwater with reduced solids because of maximizing condensate flow. For new facilities with 100% makeup, partial demineralization (e.g., reverse osmosis) may be needed to satisfy steam and boiler limits. Low pressure boilers frequently use feedwater that is suitable for use in higher pressure boilers. In these cases, the boiler water chemistry targets should be based on the pressure range that is most consistent with the makeup and feedwater quality. With local heat fluxes >150,000 Btu/hr/ft^2 (>473.2 kW/m^2), use values for at least the next higher pressure range. See Sections 3.4 and 5.1 regarding blowdown and makeup, respectively.

2. Values in the table assume the existence of a deaerator.

3. Boilers below 900 psig (6.21 MPa) with large furnaces, large steam release space, and internal chelant, polymer, and/or antifoam treatment can sometimes tolerate higher concentrations of feedwater impurities than those in the table and still achieve adequate deposition control and steam purity. Removal of impurities by external pretreatment is always a preferred solution. Alternatives must be evaluated as to practicality and economics in each individual case.

4. Nonvolatile TOC includes oily matter and any form of nonvolatile organic carbon not intentionally added as part of the water treatment regime. See Sections 4.9 and 4.10.

5. Some boiler manufacturers and technical societies recommend removing a boiler from service if pH cannot be kept above 8.0 at 25°C. If alkalinity titrations are performed rather than pH monitoring, the boiler should be removed from service if no P-alkalinity is detected. Keep in mind that titrations such as P-alkalinity cannot define the severity of pH depression, i.e., pH 8 vs pH 4, so a properly measured pH value is the better option. Operating ranges should be well above these minimums during normal operation.

Minimum and maximum levels of pH, P alkalinity or hydroxide alkalinity must be individually tailored to a particular boiler - preferably by experienced water treatment personnel with regard to silica solubility, corrosion, and other components of internal treatment. All boilers with a maximum operating pressure above 900-1000 psig should have high purity makeup and should rely primarily on pH monitoring and control rather than alkalinity. See Sections 3.2 and 4.2-4.5.

6. The total alkalinity limits in the boiler water were set at <20% of the TDS to avoid foaming in systems operating at high TDS (e.g., >250 µS/cm). If lower conductivities are used, alkalinities should be adjusted accordingly. Alkalinity targets may be adjusted if steam purity targets are met.

7. A small amount of total alkalinity will be present and measurable from treatment chemicals applied with the standard phosphate or volatile treatment programs used at these high pressure ranges (e.g., sodium phosphate, ammonia, amines, or traces of sodium hydroxide).

8. Targets for boiler water specific conductivity are set based on the maximum continuous rating (MCR) of the boiler. The specific conductivity targets are designed to limit total dissolved solids in the steam. The actual operating limit for boiler water specific conductivity should be adjusted downward from maximum values as required to ensure satisfying actual steam purity limits for the particular boiler based on steam purity monitoring and carryover studies. For example, for a 601-750 psig boiler targeting a sodium concentration of <20 ppb sodium as Na (0.020 ppm as Na) in the steam (instead of the value of 100 ppb in Table 1), the 0.05% carryover rate would translate to a maximum boiler water sodium limit of 40 ppm as Na (0.020 ppm / 0.050%=40) or about 120 ppm of TDS (40 ppm Na x 3 ppm TDS/ppm Na = 120 ppm TDS). Using a factor of 0.65 ppm of TDS per µS/cm of specific conductivity, the estimated maximum boiler water specific conductivity (120 ppm /0.65 ppm/µS/cm = 184.6 µS/cm) is approximately 180 µS/cm (See Section 3.1).

Higher boiler water conductivities may be allowable for some boilers if carryover studies demonstrate the targeted steam purity limit can still be achieved. Conversion from ppm (mg/L) TDS values in the ABMA standards [21] to our table used a factor of 0.65. See Section 4.1. For boilers operating up to 750 psig (5.17 MPa), if the boiler water TDS/conductivity factor is actually 0.5 rather than the 0.65 factor used to construct this table, the corresponding maximum specific conductivity could be 1.3 times the values shown in Table 1 (i.e., the specific conductivity limits for 300 psi (2.07 MPa), 450 psi (3.10 MPa), 600 psi (4.14 MPa), and 750 psi (5.17 MPa) would be 2100 µS/cm, 1800 µS/cm, 1500 µS/cm and 1200 µS/cm, respectively. If makeup water is of high purity, condensate is normally not contaminated, and the boiler operates at less than 1000 psig (6.89 MPa) drum pressure, the boiler water specific conductivity targets used should be those shown for the 1001-1500 psig (6.9-10.34 MPa) range. ABMA data shows mechanical carryover is 0.10% for 1800 psig (12.4MPa) boilers and 0.20% for boilers operating above this pressure.

9. Achievable steam purity depends on many variables including boiler water total alkalinity and specific conductance, as well as the design of boiler steam drum internals and operating conditions. Targets for boiler water are based on maximum continuous rating (MCR) of the boiler. As discussed in Section 3.1, one boiler manufacturer now recommends saturated steam TDS below 0.1 ppm to avoid deposits and corrosion in superheaters. Also, steam turbine and/or process steam restrictions often require lower limits than 0.1 ppm (mg/L) TDS or <30 ppb sodium.

The table includes an estimated sodium value based on an assumed TDS to sodium ratio of about 3 to 1. Since the ratio can vary for different water compositions, e.g., 2:1 to 4:1, the ratio for a particular boiler preferably would be determined from a gravimetric TDS test and sodium measurement of the boiler water sample. See Section 3.1 for setting boiler water limits needed to satisfy steam sodium limits.

10. For pressures <901 psi (<6.22 MPa), boiler water silica concentration is set to avoid silica deposits in boiler. For pressures ≥901 psi (≥6.22 MPa), silica limit is set to ensure ≤20 ppb silica in the steam. Appendix A can be used to determine the boiler water silica concentration needed to achieve a particular steam silica concentration for a particular boiler.

11. Ideally the requirements for attemperation spray water quality are the same as those for steam purity. However, higher concentrations of contamination can be tolerated in the attemperation water if the superheated steam purity limits are satisfied. In all cases the spray water should be obtained from a source that is free of deposit forming chemicals (such as sodium hydroxide, sodium sulfite and sodium phosphate) and has minimal iron and copper. Only volatile alkaline material may be used to treat water used for steam attemperation. As discussed in Section 3.1, superheated steam sodium less than 20 ppb is suggested in industrial plants and many suppliers of condensing steam turbines require sodium levels that are lower (e.g., 5 ppb or less), which may require lower boiler water conductivities. Steam purity guidelines should satisfy the requirements of the turbine manufacturer and may need to consider other parameters such as cation conductivity or degassed cation conductivity. Due to the large amounts of feedwater attemperation used by some of these units, feedwater purity values may need to match steam purity values.

SUGGESTED WATER CHEMISTRY TARGETS INDUSTRIAL WATERTUBE WITHOUT SUPERHEATER

TABLE 2   PRIMARY FUEL FIRED, DRUM TYPE

Makeup water percentage: Up to 100% of feedwater

Conditions: No superheater (Turbine drives or other process limitations may require more restrictive steam purity, boiler water and feedwater targets). [However, a low pressure boiler with an external moisture separator that consistently achieves steam purity limits shown in Table 1, may supply steam to an external superheater.]

Drum Operating Pressure	psig	0-300	301-450	451-600	601-750	751-900
	(MPa)	(0-2.07)	(2.08-3.10)	(3.11-4.14)	(4.15-5.17)	(5.18-6.21)
Feedwater (1, 2, 3, 4)						
Dissolved oxygen, ppm (mg/L) O_2 - measured before chemical oxygen scavenger addition (1, 2)		< 0.007	< 0.007	< 0.007	<0.007	<0.007
Total iron, ppm (mg/L) as Fe		≤0.10	≤0.05	≤0.03	≤0.025	≤0.02
Total copper, ppm (mg/L) as Cu		≤0.05	≤0.025	≤0.02	≤0.02	≤0.015
Total hardness, ppm (mg/L) as $CaCO_3$ (3)		<0.5	≤0.3	≤0.2	≤0.2	≤0.1

Continue

pH @ 25°C	8.8-10.5	8.8-10.5	8.8-10.5	8.8-10.0	8.8-10.0
Nonvolatile TOC (including oily matter), ppm (mg/L) as C (4)	<1	<1	<1	<0.5	<0.5
Boiler Water (5, 6)					
Silica, ppm (mg/L) as SiO_2	≤150	≤90	≤40	≤30	≤20
Hydroxide Alkalinity, ppm (mg/L) as $CaCO_3$ or pH @ 25°C (5)			See Footnote 5.		
Total alkalinity, ppm (mg/L) as $CaCO_3$ (6)	<700	<600	<500	<400	<300
Specific conductance, microsiemens/cm (µS/cm) @25°C without neutralization (6)	≤5400	≤4600	≤3800	≤3100	≤2300
Saturated Steam Purity Target (7)					
TDS (maximum), ppm (mg/L)	1.0	1.0	1.0	1.0	1.0
Maximum Mechanical Carryover [Calculated from TDS in Reference 21], %	0.029	0.033	0.040	0.050	0.067

NOTES FOR TABLE 2

1. Values in the table assume the existence of a deaerator.

2. Chemical deaeration should be provided in all cases.

3. Boilers with relatively large furnaces, large steam release space and internal chelant, polymer, and/or antifoam treatment can often tolerate higher concentrations of feedwater impurities than those in the table and still achieve adequate deposition control and steam purity. Removal of impurities by external pretreatment is always a preferred solution. Alternatives must be evaluated as to practicality and economics in each individual case. The use of a dispersant and antifoam internal treatment is typical in this type of boiler operation.

4. Nonvolatile TOC is organic carbon not intentionally added as part of the water treatment program. Nonvolatile TOC includes oily matter as well as nonvolatile TOC that is not oily. See Section 4.9 and 4.10.

5. Some boiler manufacturers and technical societies recommend removing a boiler from service if pH cannot be kept above 8.0 at 25°C. If alkalinity titrations are performed rather than monitoring pH, the boiler should be removed from service if no P-alkalinity is detected. Keep in mind that titrations such as P-alkalinity cannot define the severity of pH depression, i.e., pH 8 vs pH 4, so a properly measured pH value is the better option. Operating ranges should be well above these minimums during normal operation. Minimum and maximum levels of pH, P alkalinity or hydroxide alkalinity must be individually tailored to a particular boiler - preferably by experienced water treatment personnel with regard to silica solubility, corrosion, and other components of internal treatment. See Sections 3.2 and 4.2-4.5.

6. Alkalinity and specific conductivity values are consistent with steam purity targets in the same table. Practical targets above or below tabulated values should be individually established by careful steam purity measurements. Silica solubility is dependent on pH, temperature and the levels of trace cations. In come cases, much higher silica levels may be possible in the boiler water. Higher silica levels are considered satisfactory provided they are supported by simulation software or experience and confirmed by testing.

7. This value represents a steam purity target that should be achievable if other tabulated water quality values are maintained. The target is not intended to be, nor should it be, construed to represent a boiler

performance guarantee or recommended operating limit. More liberal steam purity limits may be specified for rapid-start, compact, water tube boilers where a lesser steam purity is achievable and deemed acceptable for the application. Also see Table 4A.

SUGGESTED WATER CHEMISTRY TARGETS INDUSTRIAL FIRETUBE

TABLE 3   PRIMARY FUEL FIRED

Makeup water percentage: Up to 100% of feedwater

Conditions: No superheater, turbine drives, or process restriction on steam purity

Drum Operating	psig	0-300
Pressure	(MPa)	(0-2.07)
Feedwater (1, 2, 3, 4)		
Dissolved oxygen, ppm (mg/L) O_2 - measured before chemical oxygen scavenger addition (1, 2)		<0.007
Total iron, ppm (mg/L) as Fe		≤0.1
Total copper, ppm (mg/L) as Cu		≤0.05
Total hardness, ppm (mg/L) as $CaCO_3$ (3)		≤1.0
pH @ 25°C		8.3-10.5
Nonvolatile TOC (including oily matter), ppm (mg/L) as C (4)		<1
Boiler Water (5, 6)		
Silica, ppm (mg/L) as SiO_2		≤150
Boiler Water pH@25 °C (5)		See discussion and footnote (5)
Total alkalinity, ppm (mg/L) as $CaCO_3$		<700
Specific conductance, μmhos/cm (μS/cm) @ 25°C without neutralization (6)		≤ 5400
Saturated Steam Purity Target (7)		
TDS (maximum), ppm (mg/L)		1.0

NS = not specified

NOTES FOR TABLE 3

1. Values in the table assume the existence of a deaerator.

2. Chemical deaeration should be provided in all cases, especially if mechanical deaeration is nonexistent or inefficient.

3. Firetube boilers of conservative design treated with internal chelant, polymer, and/or antifoam can often tolerate higher concentrations of feedwater impurities than those in the table [$\leq$0.5 ppm (mg/L) Fe, $\leq$0.2 ppm (mg/L) Cu, $\leq$10 ppm (mg/L) total hardness] and still achieve adequate deposition control and steam purity. Removal of these impurities by external pretreatment is always a preferred solution. Alternatives must be evaluated as to practicality and economics in each individual case.

4. Nonvolatile TOC is organic carbon not intentionally added as part of the water treatment program. See Section 4.9.

5. Some boiler manufacturers and technical societies recommend removing a boiler from service if pH cannot be kept above 8.0 at 25°C. If alkalinity titrations are performed rather than monitoring pH, the boiler should be removed from service if no P-alkalinity is detected. Keep in mind that titrations such as P-alkalinity cannot define the severity of pH depression, i.e., pH 8 vs pH 4, so a properly measured pH value is the better option. Operating ranges should be well above these minimums during normal operation. Minimum and maximum levels of pH, P alkalinity or hydroxide alkalinity must be individually tailored to a particular boiler - preferably by experienced water treatment personnel with regard to silica solubility, corrosion, and other components of internal treatment. See Sections 3.2 and 4.2-4.5.

6. Alkalinity and specific conductivity guidelines are consistent with steam purity target. Practical targets above or below tabulated values should be individually established for each case by careful steam purity measurements.

7. Target value represents steam purity that should be achievable if other tabulated water quality values are maintained. The target is not intended to be, nor should it be, construed to represent a boiler performance guarantee.

SUGGESTED WATER CHEMISTRY TARGETS

INDUSTRIAL, COIL TYPE, WATERTUBE, PRIMARY FUEL FIRED

FORCED CIRCULATION STEAM GENERATORS WITH INTERNAL RECIRCULATION PUMP

TABLE 4A COIL TYPE WITH INTERNAL RECIRCULATION PUMP

Makeup water percentage: Up to 100% of Feedwater
These boilers typically have coils, a steam separation drum, and a recirculation pump to provide forced water circulation through the coils. In some designs steam generation occurs in vertical watertubes rather than coil shaped tubes.

Drum Operating Pressure	psig (MPa)	0-300 (0-2.07)	301-450 (2.08-3.10)	451-600 (3) (3.11-4.14)
Feedwater (1, 2, 3, 4)				
Dissolved oxygen, ppm (mg/L) O_2 - measured after chemical oxygen scavenger addition (2)		<0.007	<0.007	<0.007
Total iron, ppm (mg/L) as Fe		≤0.1	≤0.05	≤0.03
Total copper, ppm (mg/L) as Cu		≤0.05	≤0.025	≤0.02
Total hardness, ppm (mg/L) as $CaCO_3$		<0.5	≤0.3	≤0.2
pH @ 25°C		8.8-10.5	8.8-10.5	8.8-10.5
Recirculation Pump Sample or Water to Coil (1, 2, 3, 4)				
Total alkalinity, ppm (mg/L) as $CaCO_3$		<800	<600	<500
Maximum Hydroxide alkalinity, ppm (mg/L) as $CaCO_3$ (4)		<600	<500	<400
Minimum Hydroxide Alkalinity		N.S.(4)	N.S.(4)	N.S.(4)
Silica, ppm (mg/L) as SiO_2		≤150	≤100	≤60
Specific conductance, μmhos/cm (μS/cm) @ 25°C without neutralization		<8000	<5000	<4000

Continue

Steam Purity Targets (5, 6)

Specific conductance, µmhos/cm (µS/cm) @ 25°C (5)	≤40	≤25	≤20
Total Dissolved solids, ppm (mg/L)	≤26	≤16	≤13

N.S. Not Specified, Requires evaluation by qualified personnel.

NOTES TO TABLE 4A

1. Feedwater is defined as softened makeup plus condensate entering the boiler. Water in the coil discharge is separated from steam in a separator drum. Water to coil can be sampled from the drop leg to the circulation pump or from the discharge of the coil recirculation pump. The recirculation pump or water to coil limits were based on a design in which the feedwater is fed into and mixed with the water in the separator drum, but they can also be used for units which add feedwater after the recirculation pump sample point. As this is a forced circulation boiler, the recirculation pump or water to coil is the boiler water. Where manufacturer limits are more stringent, they should be followed. For example, one manufacturer specifies zero hardness in the feedwater.

2. Chemical deaeration with catalyzed oxygen scavenger is necessary in all cases because feedwater temperature limits imposed by manufacturers of coil type watertube steam generators preclude efficient mechanical deaeration. Coil type watertube steam generators are usually treated with higher concentrations of oxygen scavenger, typical of or higher than scavenger concentrations employed for drum type boilers, to minimize failure from oxygen corrosion. Feed of chemical oxygen scavenger must be sufficient to ensure complete oxygen removal from the system before water enters the coils. Rigorous layup practices are required to minimize the potential for oxygen corrosion associated with layup and cycling service. Follow the manufacturer's recommendations for either wet or dry lay-up during idle periods. Additional information regarding lay-up and cycling service can be found in Reference 8 and 68. The latter is specific to these boiler system types and associated procedural practices to avoid coil pitting and failures.

3. A consensus for chemistry limits was not reached above 600 psig (4.14 MPa), due to the relative absence of coil type boiler operating

experience above this pressure and unresolved questions regarding appropriate or expected steam purity.

4. The total alkalinity in the low pressure category is based on manufacturer's recommendations and is more stringent than the value calculated from the recirculation pump in the steam separator tank sample. Boiler water treatment chemicals should preferably be fed to the feedwater tank to minimize sludge deposits in the coils. Hydroxide alkalinity should be maintained in the water to coil. Table 4A values are based on using softened makeup. In such systems, much of the hydroxide alkalinity is formed from decomposition of alkalinity in the makeup. The maximum hydroxide alkalinities in the table are based upon experience in which 80% of the total alkalinity decomposes to hydroxide. Hydroxide alkalinity should be maintained at a sufficient concentration (e.g., $\geq$2.5-3.0 ppm $CaCO_3$/ppm SiO_2) to help keep silica soluble and avoid complex silicate deposits. Silica solubility is dependent on pH, temperature and the levels of trace cations. In some cases, much higher silica levels may be possible in the boiler water with certain polymers designed to inhibit silica deposition. Higher silica levels are considered satisfactory provided they are supported by simulation software or experience and confirmed by testing. See Section 4.5 & Appendix A for additional guidance on silica control. Free hydroxide also helps to fluidize certain types of sludge in coil/ separator water. The maximum hydroxide alkalinities in Tables 4A and 4B are higher than for other boilers in this guide because they assume that the manufacturers provide higher forced circulation rates than achieved in natural circulation boilers. If forced circulation rates are low, much lower hydroxide levels may be required.

5. In many designs, there is no sample of the water separated from the steam before it mixes with feedwater and steam purity values assume that TDS is essentially the same on the inlet and outlet of the coil, which requires high coil circulation ratios. Depending upon the manufacturer and the steam generator model; the steam separation drums may or may not have additional internals to aid in steam separation. Due to the diversity of design, the tabulated values are based on 0.5% moisture (i.e., 0.5% mechanical carryover; fractional carry over (FCO) value of 0.005), which has been achieved by some moisture separators. These assume no superheater or turbine drive. The target is not intended to be, nor should it be, construed to represent a boiler performance guarantee.

6. Boiler antifoams are frequently used to improve steam purity. Otherwise, a reduction in alkalinity may be required. This can be achieved by increased blowdown or upgrading the pretreatment of the make-up water (e.g., by dealkalization, demineralization or reverse osmosis). However, much more restrictive chemistry limits would be required if the makeup is treated by demineralization or reverse osmosis.

SUGGESTED WATER CHEMISTRY TARGETS

INDUSTRIAL, COIL TYPE, WATERTUBE, PRIMARY FUEL FIRED

FORCED CIRCULATION STEAM GENERATORS USING FEEDWATER PUMP FOR FORCED CIRCULATION

TABLE 4B COIL TYPE USING FEEDWATER PUMP FOR FORCED CIRCULATION

Makeup water percentage: Up to 100% of water to the coil minus water recycled to the feedwater tank.

These boilers usually have coils and a centrifugal steam separator. A portion of the steam separator blowdown (trap discharge) flows to drain and a portion flows to the feedwater tank. Since the feedwater pumps force this mixture of separator blowdown, condensate and makeup through the coils, the feedwater and the *water to coil* are essentially synonymous, but the sample point is normally called feedwater.

Drum Operating Pressure	psig (MPa)	0-300	301-450	451-600 (3)
		(0-2.07)	(2.08-3.10)	(3.11-4.14)
Feedwater [Water to Coil] (1,2,3,4)				
Dissolved oxygen, ppm (mg/L) O_2 - measured after chemical oxygen scavenger addition (2)		<0.007	<0.007	<0.007
Total iron, ppm (mg/L) as Fe		≤0.6	≤0.3	≤0.18

Continue

44

Total copper, ppm (mg/L) as Cu	≤0.30	≤0.15	≤0.12
Total hardness, ppm (mg/L) as $CaCO_3$	<3.0	≤1.8	≤1.2
pH @ 25°C	10.5-12.1	10.5-12.0	10.5-11.9
Steam Separator Blowdown (Trap Discharge) (1,2,3,4)			
Total alkalinity, ppm (mg/L) as $CaCO_3$	<800	<600	<500
Maximum Hydroxide alkalinity, ppm (mg/L) as $CaCO_3$ (4)	<600	<500	<400
Minimum Hydroxide Alkalinity, ppm mg/L as $CaCO_3$ (4)	N.S. (4)	N.S. (4)	N.S. (4)
Silica, ppm (mg/L) as SiO_2	≤150	≤100	≤60
Specific conductance, μmhos/cm (μS/cm) @ 25°C without neutralization	<8000	<5000	<4000
Steam Purity Targets (5, 6)			
Specific conductance, μmhos/cm (μS/cm) @ 25°C (5)	≤40	≤25	≤20
Total Dissolved solids, ppm (mg/L)	≤26	≤16	≤13

N.S. Not Specified, Requires evaluation by qualified personnel

NOTES TO TABLE 4B

1. The steam separator typically has a trap to remove liquid removed from the steam. The steam separator blowdown is sampled from the separator drain water (i.e., concentrated boiler water). Since traps are often used, the sample is often called the trap discharge. A portion of the separator drain discharges to the feedwater tank and mixes with the softened make-up water and condensate return water. The feedwater is sampled after the feedwater tank (hotwell) or feedwater pump. Since most of this water has already been concentrated in the boiler, feedwater iron, copper and hardness limits were set six times higher for Table 4B than for Table 4A. Where manufacturer limits are more stringent, they

should be followed. For example, one manufacturer specifies zero hardness in the feedwater. The feedwater pH values or alkalinities are controlled primarily by the trapping rate and the amount of blowdown (separator discharge trap flow that is diverted to drain rather than being recycled to the feedwater tank). The frequency at which the trap opens is called the "trapping rate". For a given boiler design, the trapping rate will be greater for systems with a deaerator (e.g., 10-12 times per hour on high fire at a feedwater temperature of 250-350 °F) as compared to those using a hotwell operating below boiling (e.g., 4 to 6 times per hour on high fire at feedwater temperature of 180-190 °F feedwater). In order to avoid boiler water foaming and carryover into the steam, blowdown rate is adjusted to meet the specific conductivity and total alkalinity limits in the steam separator blowdown.

2. Chemical deaeration with catalyzed oxygen scavenger is necessary in all cases because feedwater temperature limits imposed by manufacturers of coil type steam generators preclude efficient mechanical deaeration. Coil type steam generators are usually treated with higher concentrations of oxygen scavenger, typical of or higher than scavenger concentrations employed for drum type boilers, to minimize failure from oxygen corrosion. Feed of chemical oxygen scavenger must be sufficient to ensure complete oxygen removal from the system before water enters the coils. Rigorous layup practices are required to minimize the potential for oxygen corrosion associated with layup and cycling service. Follow the manufacturer's recommendations for either wet or dry lay-up during idle periods. Additional information regarding lay-up and cycling service can be found in Reference 8 and 68. The latter is specific to these boiler system types and associated procedural practices to avoid coil pitting and failures.

3. Due to the relative absence of coil type boiler operating experience above 600 psig (4.14 MPa) and unresolved questions regarding appropriate or expected steam purity, a consensus for chemistry limits was not reached above this pressure.

4. Treatment chemical should preferably be fed to the feedwater tank to minimize sludge deposits in the coils. Hydroxide alkalinity should be maintained in the steam separator blowdown sample. Table 4A values are based on using softened makeup. In such systems, much of the hydroxide alkalinity is formed from decomposition of alkalinity in the makeup. Hydroxide alkalinity

should be maintained at a sufficient concentration (e.g., 2.5-3.0 ppm $CaCO_3$/ppm SiO_2) to help keep silica soluble and avoid complex silicate deposits. Silica solubility is dependent on pH, temperature and the levels of trace cations. In some cases, much higher silica levels may be possible in the boiler water with certain polymers designed to inhibit silica deposition. Higher silica levels are considered satisfactory provided they are supported by simulation software or experience and confirmed by testing. See Section 4.5 & Appendix A for additional guidance on silica control. Free hydroxide can also help to fluidize certain types of sludge in coil/separator water. The maximum hydroxide alkalinities in Tables 4A and 4B are higher than for other boilers in this guide because they assume that the manufacturers provide higher forced circulation rates than achieved in natural circulation boilers. If forced circulation rates are low, much lower hydroxide levels may be required.

5. Due to diversity of manufacturer designs, the tabulated values are based on 0.5% moisture (i.e., 0.5% mechanical carryover; fractional carry over (FCO) value of 0.005), which has been achieved by some moisture separators. These assume no superheater or turbine drive. The target is not intended to be, nor should it be, construed to represent a boiler performance guarantee.

6. Boiler antifoams are frequently used to improve steam purity. Otherwise, a reduction in alkalinity is required by increased blowdown or upgrading the pretreatment of the make-up water (e.g., by dealkalization, demineralization or reverse osmosis). However, much more restrictive chemistry limits would be required if the makeup is treated by demineralization or reverse osmosis.

SUGGESTED WATER CHEMISTRY TARGETS MARINE PROPULSION, WATERTUBE,

TABLE 5 OIL FIRED DRUM TYPE

Makeup water percentage: Up to 5% of feedwater

Pretreatment: At sea, evaporator condensate or better; in port, evaporator condensate or better water from shore facilities also meeting feedwater quality guidelines

Drum Operating Pressure	psig (MPa)	450-850 (3.10-5.86)	851-1250 (5.87-8.62)
Feedwater (1)			
Dissolved oxygen, ppm (mg/L) O_2 - measured before chemical oxygen scavenger addition (2)		< 0.007	<0.007
Total iron, ppm (mg/L) as Fe		≤0.02	≤0.01
Total copper, ppm (mg/L) as Cu		≤0.01	≤0.005
Total hardness, ppm (mg/L) as $CaCO_3$		≤0.1	≤0.05
pH @ 25°C		8.8-9.2	8.8-9.2
Chemicals for pre-boiler system protection		VAM	VAM
Oily matter, ppm (mg/L)		<0.05	<0.05
Boiler Water (2, 3)			
Silica, ppm (mg/L) as SiO_2		≤9.2	≤3.1
Normal Operating pH or Alkalinity Targets		N.S. (3)	N.S. (3)
Maximum Free OH alkalinity, ppm (mg/L) as $CaCO_3$)		<14 (3)	< 1 (3)
Total alkalinity, ppm (mg/L) as $CaCO_3$		N.S. (3)	N.S. (3)
Specific conductance, µmhos/cm (µS/cm) @ 25°C without neutralization (4)		<70	<50
Steam Purity (4)			
Total Dissolved Solids, ppb (µg/L)		≤30	≤30
Sodium, ppb (µg/L) as Na		≤10	≤10
Silica, ppb (µg/L) as SiO_2		≤20	≤20

VAM = Use only volatile alkaline materials.

N.S. Not Specified, Requires evaluation by qualified personnel.

NOTES TO TABLE 5

1. Values in the table assume the existence of a deaerator.

2. Some boiler manufacturers and technical societies recommend removing a boiler from service if pH cannot be kept above 8.0 at 25°C. If alkalinity titrations are performed rather than monitoring pH, the boiler should be removed from service if no P-alkalinity is detected. Keep in mind that titrations such as P-alkalinity cannot define the severity of pH depression, i.e., pH 8 vs pH 4, so a properly measured pH value is the better option. Operating ranges should be well above these minimums during normal operation. For the 450-850 psig (3.10-5.86 MPa) boilers, the maximum hydroxide alkalinity (14 ppm $CaCO_3$) produces the maximum specific conductivity (70 µS/cm) without other solids being present. Minimum and maximum levels of pH, P alkalinity or hydroxide alkalinity must be individually tailored to a particular boiler - preferably by experienced water treatment personnel with regard to silica solubility, corrosion, and other components of internal treatment. All boilers with a maximum operating pressure above 900-1000 psig should have high purity makeup and should rely primarily on pH monitoring and control rather than alkalinity. See Sections 3.2 and 4.2-4.5.

3. Suggested maximum boiler water specific conductivity values are intended to serve as an alarm for salt water condenser leaks and can be correlated with chloride ion content in feedwater and/ or boiler water. The maximum value may need to be reduced to comply with steam purity targets.

4. Carryover testing is required as some boilers may require much lower boiler water conductivities and alkalinities to achieve desired steam purity. The targets are not intended to be, nor should they be construed to represent, boiler performance guarantees.

SUGGESTED WATER CHEMISTRY TARGETS ELECTRODE, HIGH VOLTAGE,

TABLE 6 FORCED CIRCULATION JET TYPE

Makeup water percentage: Up to 100% of feedwater

Conditions: No superheater, turbine drives, or process restriction on steam purity

Operating Pressure	psig	0-450
	(MPa)	0-3.1
Feedwater (1, 2, 3, 4)		
Dissolved oxygen, ppm (mg/L) O_2 – measured before chemical oxygen scavenger addition (1)		<0.007
Total hardness, ppm (mg/L) as $CaCO_3$ (2)		≤0.25
pH @ 25°C		8.5-10.5
Nonvolatile TOC, ppm (mg/L) as C (3)		N.S.(4)
Boiler Water (5, 6, 7, 8, 9)		
pH @ 25°C		8.5-10.5 (5)
Hydroxide alkalinity, ppm (mg/L) as $CaCO_3$ (4)		(5)
Silica, ppm (mg/L) as SiO_2		≤150
Total alkalinity, ppm (mg/L) as $CaCO_3$		<350 (5, 6)
Chloride, ppm as Cl		<300 ppm (7)
Total iron, ppm (mg/L) as Fe plus total copper, ppm (mg/L) as Cu		≤2.0 (2, 8)
Suspended solids		N.S.(8)
Organic matter		N.S.(4)
Specific conductance, μmhos/cm (μS/cm) @ 25°C without neutralization		N.S.(9)
Steam Purity		N.S.

N.S. Not Specified

NOTES TO TABLE 6

1. Values in the table assume the existence of a mechanical deaerator. Chemical deaeration should be provided in all cases, especially if mechanical deaeration is nonexistent or inefficient.

2. Some boilers may tolerate higher concentrations of feedwater impurities than those in the table and still achieve adequate deposition control.

3. Nonvolatile TOC is organic carbon not intentionally added as part of the water treatment program. See Section 4.9.

4. Naturally occurring organics, particularly when combined with hydroxide alkalinity, may cause foaming of the boiler water. Ground fault arcing between the electrode and upper boiler shell may result.

5. Free hydroxide alkalinity concentrations are not specified for jet type electrode boilers, Table 6, because the targets vary depending on the manufacturer, materials of construction, and boiler design. BSI 2486, 1997, advised a minimum boiler water pH of 9.5 [43]. The exceedingly high recirculation in these boilers creates a high potential for foaming, especially where organic contamination of feedwater might occur. Maximum hydroxide alkalinity concentration must be individually specified by experienced water treatment personnel with regard to silica solubility, corrosion, organic matter concentration, and other components of internal treatment. See Sections 3.2 and 4.2-4.5.

6. The use of high alumina porcelain insulators may allow the target total alkalinity to be increased to 600 ppm (mg/L) as $CaCO_3$.

7. This target chloride concentration is required if the electrode boiler contains stainless steel components. The exact limit depends on the type of alloy.

8. Suspended solids present in the boiler water contribute to erosion/corrosion of the electrodes and counter electrodes. Additionally, the presence of suspended solids in the boiler water increases the potential for foaming and ground fault arcing. ABMA, (Table 9 in 2012 Guide) limits for Electrode Type Boilers specifies iron levels to <0.5 ppm [21].

9. Boiler performance is determined by the specific conductivity of the boiler water. One supplier advised <3500 µS/cm. The optimum specific conductivity range is dependent on the specific boiler design. Consult manufacturer's guidelines.

SUGGESTED WATER CHEMISTRY TARGETS PROCESS WASTE HEAT BOILERS PROVIDING SUPERHEATED STEAM TO CONDENSING STEAM TURBINES AND TURBINE DRIVES

TABLE 7

[Examples of processes include ammonia, ethylene and methanol production, etc. Process waste heat boilers also include Transfer Line Exchangers (TLEs).]

Makeup water percentage: Up to 100% of feedwater

Makeup type: High purity water advised

Conditions: Supplies superheated steam to condensing steam turbines and turbine drives

High duty and operating pressure of 1000 to 1800 psig (6.9-12.4 MPa)

For lower pressure waste heat boilers see Note 1.

Operating Pressure	psig (MPa) 6.9-12.4	1000-1800
Feedwater (1, 2)		
Dissolved oxygen, ppm (mg/L) O_2		<0.007
pH at 25°C (copper alloys present)		8.8 to 9.2
pH at 25°C (no copper alloys present)		9.4-9.8
Silica, ppb (µg/L) as SiO_2 (2)		≤20
Iron, ppb (µg/L) as Fe		≤10
Copper, ppb (µg/L) as Cu		≤3
Boiler Water (1, 3, 4, 5)		
Silica, ppm (mg/L) as SiO_2		≤1.0 (3)
pH and Phosphate		(See 4, 5)
Hydroxide Alkalinity, ppm (mg/L) as $CaCO_3$		<2 (4, 5)
Maximum Total Alkalinity, ppm (mg/L) as $CaCO_3$		N.S. (4)
Specific Conductivity, µS/cm at 25°C		≤50

Continue

Steam (6)	
TDS (maximum), ppm (mg/L)	≤0.03
Sodium, ppb (µg/L) as Na	≤10
Maximum Mechanical Carryover [Calculated from TDS in Ref. 21], %	0.1%
Silica, ppb (µg/L) as SiO_2	≤20

N.S. Not Specified

NOTES TO TABLE 7

1. For low duty and/or low pressure process waste heat boilers, see feedwater and boiler water criteria in Tables 1, 2 and 3. If the makeup and condensate are sufficiently pure to satisfy the limits shown, the chemistry guidelines for higher duty process waste heat boilers in Table 7 may be used, although the silica limit probably should be adjusted using the methodology in Appendix A.

2. The feedwater silica target is more stringent than required for steam purity. This target is designed to minimize other contamination that may be present along with the silica.

3. Boiler water silica limit is 1.0 ppm at the maximum pressure (1800 psig) and 5.9 ppm at the minimum drum pressure (1000 psig) for 20 ppb silica in the steam. See Appendix A for other pressures as illustrated in Table 8.

4. At least one supplier of these boilers says circulation is not sufficient to support phosphate treatment and all volatile treatment is required. If using all volatile treatment, sodium balancing (54 ppb as Na of sodium per µS/cm of cation conductivity) is suggested if cation conductivities exceed 1.5 µS/cm in the boiler water [46]. However, phosphate treatment is recommended for the boiler water, provided that it is consistent with manufacturer recommendations and phosphate hideout is not excessive (see footnote 5). Consider consulting a chemical treatment consultant or contractor to develop a suitable treatment program. While normal operating pH should be noticeably higher (e.g., >9.0), remove boiler from service if pH cannot be kept above 8.0 at 25°C. Total alkalinity is not monitored and is generally not applicable to boilers operating with such low levels of solids in the boiler water.

5. A consensus was not reached on a type of phosphate treatment program for these boilers. One author recommended 1-3 ppm as PO_4 and a minimum sodium to phosphate mole ratio of 2.8:1 for transfer line exchangers (TLEs)[73]. Some boilers with good circulation and low heat flux may operate with higher phosphate concentrations (3-6 ppm as PO_4 in the higher pressure range or 4-10 ppm as PO_4 in the lower pressure range) and control pH within a band of congruent sodium to phosphate mole ratios calculated from pH, phosphate, and ammonia and amines. Methods for correcting ratios are presented in References [49, 71, 72]. Sodium to phosphate mole ratios after correcting for the effect of ammonia and amines should be >2.2:1 to 3.0:1 and often requires more restrictive limits [72]. Some process boilers may experience significant phosphate hideout (see glossary) requiring equilibrium phosphate treatment [See 41] in which the maximum phosphate concentration allowed is ~2.5 ppm as PO_4 or less. In such cases, the maximum phosphate limit is set at or below the concentration at which phosphate hideout is experienced, the solid alkali pH is kept at ≥9.0 at 25°C and free hydroxide should be kept less than 2 ppm as $CaCO_3$ (and more typically ≤1.0 ppm as NaOH). If the maximum phosphate residual without hideout is less than 1.0 ppm as PO_4, concerns typically associated with all volatile treatment may begin to apply and it is suggested to seek expert assistance and consider using more advanced chemistry monitoring techniques such as cation conductivity and sodium in the boiler water.

6. Many suppliers of condensing steam turbines require sodium levels that are lower (e.g., 5 ppb or less) which may require lower boiler water conductivities. Steam purity guidelines should satisfy the requirements of the turbine manufacturer and may need to consider other parameters such as cation conductivity or degassed cation conductivity. Due to the large amounts of feedwater attemperation used by some of these units, feedwater purity values may need to match steam purity values. The mechanical carryover rates are based on ABMA data. Units made in other countries may have different mechanical carryover rates.

SECTION 6: REFERENCES

1. American Society of Mechanical Engineers. Consensus on Operating Practices for the Control of Feedwater and Boiler Water Quality in Modern Industrial Boilers, ("Blue Book"), 1979. This version is out of print.

2. American Society of Mechanical Engineers. Consensus on Operating Practices for the Control of Feedwater and Boiler Water Quality in Modern Industrial Boilers, CRTD-Vol. 34.,1994. This version is out of print.

3. American Society of Mechanical Engineers. Consensus on the Operating Practices for the Control of Feedwater and Boiler Water Chemistry in Modern Industrial Boilers," CRTD.Vol.34.1998. ISBN 0-7918-1204-9. With Errata page issued in January 2001.

4. American Society of Mechanical Engineers, Consensus on Operating Practices for Control of Water and Steam Chemistry in Combined Cycle and Cogeneration Power Plants, CRTD ASME Order No. 859988, 2012, ISBN 978-07918-5998-8.

5. American Society of Mechanical Engineers, Consensus on Pre-Commissioning Stages for Cogeneration and Combined Cycle Power Plants, 2017,CRTD ASME Order No. 802485, 2017, 345 East 47th Street, New York, New York, 10017 U.S.A.ISBN 978-07918-6126-4.

6. American Society of Mechanical Engineers, A Practical Guide to Avoiding Steam Purity Problems in the Industrial Plant, CRTD-Vol. 35, 1995, ISBN 0-7918-1220-0.

7. American Society of Mechanical Engineers, Consensus on Operating Practices for the Sampling and Monitoring of Feedwater and Boiler Water Chemistry in Modern Industrial Boilers, CRTD-Vol. 81, 2006, ISBN 0-7918-0248-5.

8. American Society of Mechanical Engineers, Consensus for the Layup of Boilers, Turbines, Turbine Condensers and Auxiliary Equipment," CRTD-Vol. 66, 2002, ISBN 0-7918-3618-5.

9. American Society of Mechanical Engineers, Consensus on Best Tube Sampling Practices for Boilers and Nonnuclear Steam Generators, CRTD-Volume 103, 2014, Three Park Avenue, New York, NY 10016, USA. ISBN 9780791860359.

10. American Society of Mechanical Engineers, Deaerators, Performance Test Codes, ANSI/ASME PTC 12.3-1997, ASME Order No. D02397, ISBN 0-7918-2454-3.

11. American Society of Mechanical Engineers, High Purity Water Treatment Systems, Performance Test Code PTC 31-2011, ASME Order No. C01611, ISBN 9780791834008.

12. Daniels, G. C. Prevention of turbine-blade deposits. ASME Paper 48-SA-25. Abstracted in Mech. Eng. 70:694-95, 1948.

13. Coulter; E. E., E. A. Pirsh, and E. J. Wagner, Jr. Selective silica carryover in steam. Trans. ASME, 78:869-873, 1956.

14. Jonas, O. and B. C. Syrett. Chemical transport and turbine corrosion in phosphate treated drum boiler units. Proc. Int'l Water Conf., Eng. Soc. W. Pa., 48, 158-166, 1987.

15. Jonas, O. Determination of steam purity limits for industrial turbines. Proc. Int'l Water Conf., Eng. Soc. W. Pa., 49, 137-147, 1988.

16. Dewitt-Dick, D., J. S. Beecher, and F. Seels. Steam purity problems encountered in industrial turbines. Proc. Int'l Water Conf., Eng. Soc. W. Pa., 49, 148-159, 1988.

17. Whitehead, A. and R. T. Bievenue. Steam purity for industrial turbines, Proc. Int'l Water Conf., Eng. Soc. W. Pa., 49, 160-172, 1988.

18. Navitsky, G. and H. A. Grabowski. Steam purity for industrial steam generators. Proc. Int'l Water Conf., Eng. Soc. W. Pa., 49, 173-180, 1988.

19. American Society of Mechanical Engineers. The ASME Handbook on Water Technology for Thermal Power Systems. New York, 1989.

20. Fynsk, A. and J. O. Robinson. A practical guide to avoiding steam purity problems in the industrial plant. Proc. Int'l Water Conf., Eng. Soc. W. Pa., 53, 415-425, 1992.

21. American Boiler Manufacturers Association (ABMA). Boiler Water limits and Achievable Steam Purity for Water Tube Boilers,

4th edition, Arlington, VA. 1995. Boiler Water Quality Requirements and Associated Steam Quality for Industrial/Commercial and Institutional Boilers, 2005, Vienna, VA. Also See Boiler Water Quality Requirements and Associated Steam Quality for ICI Boilers, 2012, Vienna, VA.

22. Bellows, J. C., "Steam Purity Recommendations for New Small Steam Turbines", International Water Conference 1995, Paper IWC-95-19.

23. Harvey A.H. and Bellows, J.C., "Evaluation and Correlation of Steam Solubility Data for Salts and Minerals of Interest in the Power Industry, NIST Technical Note 1287 (U.S. Government Printing Office: Washington, D.C. 1997).

24. The Babcock & Wilcox Company, Section 42, Water and Steam Chemistry, Deposits and Corrosion, Steam, Its Generation and Use, Edition: 41, Barberton, OH, 2005.

25. Banweg, Anton "Industrial Steam Purity: Requirements, Proper Sampling and Practical Considerations," International Water Conference 2008, Paper IWC-08-28.

26. Beardwood, Edward S., Silica in Steam Generating Systems, IWC-08-12, International Water Conference, Engineers Society of Western Pennsylvania, 2008.

27. Bellows, James. Sodium Hydroxide in Steam Turbines, IWC-08-14, International Water Conference, Engineers Society of Western PA, 2008.

28. Bellows, James C., Mass Transfer and Deposition of Impurities in Steam Turbines, PowerPlant Chemistry, Volume 10, No. 2, 2008, pp. 118-122.

29. IAPWS, TGD1-08, Technical Guidance Document: Procedures for the Measurement of Carryover of Boiler Water into Steam (2008). Available from http://www.iapws.org.

30. IAPWS, TGD2-09(2015), Technical Guidance Document: Instrumentation for monitoring and control of cycle chemistry for the steam-water circuits of fossil-fired and combined-cycle power plants, 2015. Available from http://www.iapws.org.

31. EPRI, Comprehensive Cycle Chemistry Guidelines for Fossil Plants, 1021767, EPRI, Palo Alto, CA, 2011.

32. IAPWS, TGD5-13, Technical Guidance Document: Steam Purity for Turbine Operation, International Association for the Properties of Water and Steam (IAPWS), 2013, Available from http://www.iapws.org.

33. Beardwood, Edward, S. "Steam Purity in Heat Recovery Steam Generators, International Water Conference, IWC-14-28, Engineers Society of Western Pennsylvania, 2014.

34. Bellows, James, Chemistry Aspects of Industrial Turbines, IWC-16-56, International Water Conference, Engineers Society of Western PA, 2016.

35. Dillon, James J., Paul B. Desch, Tammy S. Lai, The Nalco Guide to Boiler Failure Analysis, Second Edition, Daniel J. Flynn Editor, McGraw Hill, NY, NY, 2011. ISBN 9780071743006.

36. Guidelines for Controlling Flow-Accelerated Corrosion in Fossil and Combined Cycle Plants, EPRI, Palo Alto, CA: 2017. 3002011569. (Available free of charge from www.epri.com.)

37. Whirl, S. F. and T. E. Purcell. Protection against caustic embrittlement by coordinated phosphate pH control. Proc. Annual Water Conf., Eng. Soc. W. Pa., 3, 45-60B, 1942.

38. Marcy, V. M. and S. L. Halstead. Improved basis for coordinated phosphate pH control of boiler water. Combustion 35: 4547, 1964.

39. Betz, Betz Handbook of Industrial Water Conditioning, Eighth Edition, Trevose, PA, 1980.

40. McCoy, James W., The Chemical Treatment of Boiler Water, Chemical Publishing Co., NY, NY, 1984 (2nd Printing). ISBN 0820602841.

41. Stodola, Jan. Review of Alkalinity Control for Boiler Water, IWC-86-27, Proceedings of the 47th International Water Conference, Engineers Society of Western Pennsylvania, 1986.

42. Lane, Russel W., Control of Scale and Corrosion in Building Water Systems, McGraw Hill, NY, NY, 1993. ISBN 0070362173.

43. BSI. Recommendations for Treatment of water for steam boilers and water heaters, BS 2486: 1997, Health and Environmental Sector Board, February 1997.

44. Frayne, Colin. Boiler Water Treatment Principles and Practice, Volume II, Treatments, Program Design and Management, Chemical Publishing Co. Inc., New York, NY, 2002.

45. Bartholomew, Robert D., An Introduction to Alkalinity Limits for Boiler Water Treatment, Paper 08-13, International Water Conference, Engineers Society of Western Pennsylvania, 2008.

46. Bartholomew, Robert D., Sodium Balancing for Drum-Type Boilers on All Volatile Treatment, PowerPlant Chemistry, Vol. 11, No. 9, pp. 533-539, 2009.

47. IAPWS, TGD3-10(2015), Technical Guidance Document: Volatile treatments for the steam-water circuits of fossil and combined cycle/HRSG power plants, International Association for the Properties of Water and Steam (IAPWS), 2015, Available from http://www.iapws.org.

48. IAPWS, TGD4-11(2015), Technical Guidance Document: Phosphate and NaOH treatments for the steam-water circuits of drum boilers of fossil and combined cycle/HRSG power plants, International Association for the Properties of Water and Steam (IAPWS), 2015, Available from http://www.iapws.org.

49. Bartholomew, R. D., Correct for Ammonia/Amine Effect on pH to Avoid Corrosion with Phosphate Treatment, IWC-14-26, 75th International Water Conference, Engineers Society of Western Pennsylvania, 2014.

50. IAPWS, TGD8-16(2019), "Application of Film Forming Amines in Fossil, Combined Cycle, and Biomass Power Plants," International Association for the Properties of Water and Steam, Dresden, Germany, 2019.

51. Nalco, The Nalco Water Handbook, 4th Edition, McGraw Hill, NY, NY, 2017. ISBN 9781259860973.

52. Beardwood, Edward S., Determination of the Time to Clean Industrial boilers Based Upon Upset Feedwater Conditions, IWC-19-30, International Water Conference, Engineering Society of Western Pennsylvania, 2019.

53. American Society for Testing and Materials. 1986. Designation E 380 86, Metric practice. Annual Book of ASTM Standards, Vol. 14.02, Philadelphia.

54. TAPPI, Response to contamination of high purity boiler feedwater, TIP 0416-05, 2005.

55. APHA, AWWA, WEF, 2320 Alkalinity, Standard Methods for the Examination of Water and Wastewater, 21st Edition, Port City Press, Baltimore, MD, 2005.

56. Powell, Sheppard T., "Hydroxide Alkalinity: Barium Chloride Method", Water Conditioning for Industry, McGraw-Hill Book Company, Inc., 1954, p. 475.

57. American Society for Testing and Materials. 1988. Designation D 888 87, Standard test methods for dissolved oxygen in water. Annual Book of ASTM Standards, Vol. 11.01, 462-473

58. Hitchman, M. I. Measurement of dissolved oxygen. Chemical Analysis, Vol. 49, New York: John Wiley & Sons, 1978.

59. Weick, R. H. How to determine when an industrial boiler needs cleaning. Proc. Int'l. Water Conf., Eng. Soc. W. Pa., 36, 71-76, 1975.

60. Weick, R.H. Internal Boiler Tube Deposit Weights as a Basis for Chemical Cleaning, Paper No. 208. Corrosion 94. Houston, TX: National Association of Corrosion Engineers, 1994.

61. Bloom, D., Prepared Discussion of: The Influence of System Parameters on TOC Degradation, IWC-13-75D, International Water Conference, Engineers Society of Western PA, 2013.

62. American Society for Testing and Materials. Designation D 2579 85, Method A, Standard test methods for total and organic carbon in water (oxidation and infrared detection). Annual Book of ASTM Standards, Vol. 11.02, 12-14, Philadelphia, 1985.

63. American Public Health Association. 1989. Oil and grease. Standard Methods for the Examination of Water and Wastewater, 17th ed., 541-548, Washington, D.C.

64. Association of Water Technologies, Chapter 3 Boiler Systems, Technical Manual,.

65. American Petroleum Institute, API Recommended Practice 538, Industrial Fired Boilers for General Refining and Petrochemical Service, First Edition, October 2015 (www.api.org).

66. ASTM International, D1066-18, Standard Practice for Sampling Steam,2019 Annual Book of ASTM Standards, Volume 11.01 Water (1), West Conshohocken, PA, 2019.

67. National Institute of Standards and Technology, NIST Guide for the Use of the International System of Units (SI), NIST Special Publication 811 2008 Edition.

68. Beardwood, Edward S., Operational Control and Maintenance Integrity of Typical and Atypical Coil Tube Generating Systems, NACE Corrosion 99, Paper No. 338, 1999, NACE International, Houston, Texas, USA.

69. Klein, H.A., Lux, J.A., Riedel, W.L., Noll, D.E., Phillips, H., A Field Survey of Internal Boiler Tube Corrosion in High Pressure Utility Boilers, Presented at the American Power Conference, April, 1971.

70. Checkovich, Andrew, Progress Report No. IV General Study of Boiler Tube Corrosion Logan Plant, Combustion, January 1980, pp. 11-20.

71. Puchan, David G., Bryant, Robert C, Calculating the Impact of Amine Contribution to the $Na:PO_4$ Molar Ratio Calculation, IWC-20-12. International Water Conference, Engineers Society of Western Pennsylvania, 2020.

72. Bartholomew, R.D., Prepared Discussion of: Calculating the Impact of Amine Contribution to the Na:PO4 Molar Ratio Calculation, IWC-20-12D, International Water Conference, Engineers Society of Western Pennsylvania, 2020.

73. Robinson, James O., Avoiding Waterside Corrosion Problems in Ethylene Plant Steam Systems, 2014 AIChE Ethylene Producers Conference, New Orleans, LA, 2014.

GLOSSARY

Alkalinity

Alkalinity can be described as bicarbonate, carbonate, phosphate, and hydroxide. Each of these components will contribute some amount of basicity to a water solution. The sum of all alkalinity components is referred to as total alkalinity.

Boiler water

Cycled up feedwater inside the boiler. Blowdown from the steam drum of a watertube boiler and bottom or surface blowdown from a firetube boiler are typically considered representative of the water chemistry inside the boiler.

Caustic Gouging

The dissolution of carbon steel by localized high concentrations of sodium hydroxide to form sodium ferroate and sodium ferroite. Thick, black deposits typically remain in the gouged area. Corrosion also requires a concentrating mechanism, such as steam blanketing or boiling within porous deposits, and buildup of a thermal barrier from cooling water.

Cycles of Concentration

In the absence of carryover, it is the feedwater flow rate divided by blowdown flow rate. It is the extent of evaporative concentration of feedwater dissolved solids in the boiler and can be calculated by (boiler water chloride / feedwater chloride) for softened makeup. It is often estimated by (boiler water specific conductivity / feedwater specific conductivity) although some of the specific conductivity can be due to boiler water treatment chemicals or alkalinity decomposition upon entrance into boiler. Sometimes tracer chemicals are fed to feedwater and monitored in boiler water to determine cycles of concentration.

Deaerated water

The process of deaeration removes oxygen from water by either mechanical or chemical means. Mechanical deaerators that are functioning properly and operating at design conditions should reduce oxygen to ≤ 7 µg/L (ppb) oxygen as O_2. Deaerating heaters may only achieve removal to <40 µg/L (ppb) oxygen as O_2. Scrubbing/reboiler deaerating heater performance can be significantly reduced at low flow rates.

Direct spray water used for steam attemperation

Also known as attemperation water. Direct spray water is injected into superheated steam to control its temperature. Any solids present in direct spray water can potentially cause deposits and/or corrosion within the attemperator or directly downstream, so this water must be essentially solids free.

Electrode, forced circulation jet type boiler

Electric steam generators produce steam rapidly and can be taken offline very quickly. The electrode forced circulation jet type electric boiler uses the conductive and resistive properties of water to generate steam. Because water has electrical resistance, the current flow from electrodes immersed in or sprayed with boiler water generates heat directly in the water and steam production occurs.

Feedwater

Water supply going to the boiler. It is typically a combination of condensate return and water supplied from the makeup treatment system. Tables specify deaerated water, which typically is achieved by a deaerator located before the boiler feed pump(s).

Holding Time

Boiler volume divided by blowdown flow rate. This is the length of time to remove one boiler volume of water.

High purity water

Treated water that has all or most of the dissolved minerals removed. See Section 5.1 for minimum effluent purity assumed. Removal is usually by membrane separation, ion exchange (demineralization), distillation (evaporation), and combinations of these processes. Commonly called demineralized water or deionized (DI) water.

Industrial watertube boiler

Industrial watertube boilers are composed of drums and tubes. Water and steam are contained on the inside of boiler tubes, while heat is applied to the exterior of the tubes. The water provides cooling to the tube metal and prevents overheating. A multitude of boiler designs, sizes, and manufacturers exist. All forms of fuel can be burned. Steam produced can be saturated or superheated.

Industrial firetube boiler	The industrial firetube boiler consists of a shell filled with water through which hot gases are passed on the inside of tubes located within the shell. They are typically limited to less than 300 psig, produce saturated steam, and are generally smaller in size than industrial watertube boilers. Firetube boilers are limited to burning liquid and gas type fuels, although a few hybrids exist that can also burn biomass.
Industrial coil watertube steam generator	Industrial coil watertube steam generators are available from several manufacturers; each OEM having unique design characteristics. The main design feature of these generators is a heated coil or series of coils through water travels, entering as water on one end and leaving as a mixture of steam and water on the other end. The coil is followed by a steam separator generally capable of achieving steam qualities greater than 98%. Coil steam generators will typically have a low water volume to boiler horsepower ratio, small footprint, and quick steaming capability.
Limit	Values presented as limits in this document are suggested consensus values.
Marine propulsion watertube boiler	Marine propulsion refers to the mechanism or system used to generate thrust and move a ship through the water. Most modern ships are propelled by mechanical systems consisting of an electric motor or engine turning a propeller. Although the steam turbine was the source of this electricity for most marine vessels at one time, new-build ships with steam turbines are generally limited to merchant vessels where the cargo can be used as fuel (e.g., LNG or coal carriers).
Nonvolatile TOC	An unofficial modification of the TOC test suggested by this task group in which the TOC test is conducted on a sample after atmospheric boiling with the subsequent subtraction of a calculated carbon value equivalent to the carbon content of any nonvolatile organic treatment chemicals fed

Nonvolatile treatment chemical	A nonvolatile treatment chemical, such as phosphate, polymer, or sulfite, will leave solids/deposits when water in which it is dissolved is boiled to dryness.
Oily matter	Includes all nonvolatile hydrocarbons, vegetable oils, animal fats, waxes, soaps, greases, and related matter which are extractable in hexane or halogenated solvents at low pH. This grouping unfortunately excludes some potentially damaging organic feedwater contaminants and includes some beneficial organic compounds because of their solvent solubility.
Organic matter	Broad category of potential contaminants which contain one or more carbon atoms in their structure.
Phosphate Gouging	The dissolution of carbon steel by localized high concentrations of mono- or disodium phosphate to form maricite ($NaFePO_4$) and other sodium iron phosphate compounds. Crusty alternating black-and-white layered deposits may remain with knife-edge-like surfaces in the gouged area. Corrosion also requires a phosphate concentrating mechanism, such as steam blanketing or boiling within porous deposits, and buildup of a thermal barrier from cooling water.
Phosphate Hideout	Apparent loss of boiler water phosphate in high-pressure water-tube boilers operating under high load conditions. The salts reappear when the load is reduced.
Process Waste Heat Boilers	These are process industry boilers that utilize waste heat sources primarily from the refining and chemical process industries such as Transfer Line Heat Exchangers (TLE). It does not include gas turbine heat recovery steam generators.
Softened water	The term softening typically refers to the reduction of the calcium and magnesium hardness from a water source. This can be accomplished by a variety of pretreatment processes including lime softening, ion exchange, membrane separation, or evaporation. See Section 5.1 for more information.

Specific conductivity

Specific conductance is a measure of the ability of water to conduct an electrical current. It is highly dependent on the amount and nature of the dissolved solids present in the water. Pure water, such as distilled water, will have a very low specific conductance while water with a high dissolved solids concentration will have a much larger specific conductivity. Due to its general correlation with dissolved solids concentration and the ease with which it is determined, specific conductivity is often used instead of measuring TDS. Specific conductivity may also be referred to as unneutralized specific conductivity or unneutralized conductivity.

Steam purity

An expression of the quantity of non-water components contained in the steam. These components can be dissolved in the steam, dissolved in water droplets entrained in the steam, or carried as discrete solid particles with the steam.

Steam quality

Relates to the quantity of moisture present in the steam with 100% steam quality specifying no moisture content and 0% quality specifying all liquid.

Superheater

A superheater is a component of the boiler that heats steam above its saturation temperature. It has tubes with steam flow on the interior and heat on the exterior. Steam heats up as it flows through the tubes. Superheated steam has no moisture, i.e., has a steam quality of 100%, and must lose all superheat before condensation can occur.

Total dissolved solids (TDS)

Total dissolved solids are a measure of the dissolved content of all inorganic and organic substances present in a liquid in molecular and ionized form. TDS is limited in the boiler water to primarily prevent carryover, and to a lesser degree, to prevent scaling and corrosion. (Scaling and corrosion in the boiler are largely controlled by feedwater quality.) Since the TDS measurement requires boiling a known amount of water to dryness and subsequent determination of the weight of solids left behind, it cannot be automated. Specific conductivity, which can be automated and is much easier to determine, is often substituted for the TDS measurement.

Volatile
Organic
Matter

Volatile organic matter (VOM) can include both harmful and non-harmful compounds. It travels with the steam and may cause damage to turbines and other steam system equipment. Potentially harmful VOM commonly present includes low molecular weight organic acids such as formic, acetic, and glycolic.

APPENDIX A. ESTIMATING BOILER WATER SILICA LIMITS

As stated in Section 4.5, the silica limits in Tables 1-5 are set to avoid silica deposits within boilers operating at ≤900 psig. When there is a turbine manufacturer or other definitive limit for silica in the steam, it is suggested to use the following sequence for estimating an appropriate boiler water silica limit. This process was developed primarily from the cited references: Silica in Steam Generating Systems and The Chemical Treatment of Boiler Water [26 and 40].

1. Determine silica target (ppm SiO_2) in steam (Si_{target}) to the turbine or other limiting condition for the system. For nonreheat condensing turbines, the steam silica limit usually is 0.020 ppm. In noncondensing turbines, consult turbine manufacturer recommendations or solubility data [23,26,28].

2. Determine the highest proportion (%) of attemperation water (A) commonly injected into the steam and total silica (ppm SiO_2) in the attemperation water (Si_{attemp}).

3. Determine saturated steam silica target:
   $$(Si_{sat}) = (Si_{target} - (A \times Si_{attemp}))/(1-A) \text{ in ppm}$$

4. Determine the boiler water silica target to maintain steam purity. These formulas are setup to be used in spread sheets. Spread sheets automatically interpret % to be a fraction (e.g. 0.033% = 0.00033). However, when solving these equations on a calculator, the "% Mechanical Carryover" will need to be divided by 100% to cancel the percentile.

   a. For boilers with drum pressures of ≤500 psig, volatile carryover is sufficiently low and the estimates for mechanical carryover in Table 1 should be sufficiently conservative to provide a quick estimate for meeting a saturated steam silica limit.

   $Si_{boiler} = Si_{sat}$ / (% mechanical carryover).

   b. For a more conservative estimate of the boiler water silica needed to achieve a given steam silica concentration (and for all boilers operating over 500 psig), one should combine the mechanical and volatile silica carryover to obtain a total silica carryover estimate. For this calculation only, Committee advises to use half of the mechanical carryover level noted in Table 1 (to account for the boiler manufacturer's factor of safety in carryover) plus the following McCoy volatility formula. Table 8 presents the results of example calculations using this formula, 0.020 ppm of silica in the steam, and assumed pH values. Higher boiler water pH values decrease silica volatility which will reduce steam silica at the same boiler water silica concentration.

% Volatile Silica Carryover = $10 \char94 (0.0106 \times Ts - 0.17 \times (EPM_{OH})\char94 0.5 - 8.27)$.

Where:

Ts = Saturation Temperature °F.

EPM_{OH} or equivalents per million of hydroxide = $1000 \times 10\char94(pH-14)$.

Alternatively, EPM_{OH} may be estimated as 2% of the Barium chloride hydroxide alkalinity (i.e., hydroxide alkalinity in mg/L $CaCO_3/50$).

% Total Silica Carryover = %Mechanical Carryover/2 + % Volatile Silica Carryover

$Si_{boiler} = Si_{sat}$ / (% Total Silica Carryover).

5. Compare this target to the silica targets based on the boiler pressure in the appropriate table (Tables 1-7 of this guideline) and pick the lower of the two silica targets. However, since silica limits for boilers operating above 901 psi (>6.22 MPa) in Table 1 were set to achieve 20 ppb SiO_2 based on the maximum pressure in each column, tabular values may be stricter than necessary for some boilers in the lower end of each pressure range.

6. If predicted silica concentration at the maximum allowable or planned blowdown may exceed the selected target, then improve makeup and/or condensate purity.

7. When boiler water silica concentrations are normally elevated (>9 ppm), the caustic alkalinity in mg/L $CaCO_3$ should be at least 2.5 times the silica concentration [p. 413 of Ref 44]. This is consistent with other sources [43]. Beardwood set limits at 2.8 ppm $CaCO_3$ of hydroxide alkalinity per ppm of silica in boiler water. Higher silica concentrations may be possible with certain polymers designed to inhibit silica deposition.

8. Direct testing of steam silica and boiler water silica concentrations can be used to ensure that the estimated boiler water silica limits are appropriate to satisfy steam purity requirements.

For low pressure boilers without steam turbines, the boiler water silica limits can be due to factors other than steam purity such as avoiding the formation of complex silicates. Lower boiler water concentrations may be required if deposits are noted from silica concentrating under deposits or from silica combining with hardness, iron, aluminum or other cations. Where possible, makeup and condensate treatment should be improved to minimize concentrations of these cations. Higher silica concentrations than listed in the tables may be safely utilized in some cases.

EXAMPLE CALCULATIONS OF BOILER WATER SILICA LIMIT, 0% ATTEMPERATION, 0.020 PPM SATURATED STEAM SILICA

TABLE 8

Limit Table	Drum Pressure, psig	Sat. Temp., °F	Minimum pH (1)	Estimated OH Alkalinity, ppm CaCO$_3$ (2)	EPM of Hydroxide	%Mech. Carryover	% Volatile Silica Carryover	%Total Silica Carryover	Si$_{Boiler}$, ppm SiO$_2$
1	300	422	10.00	5.0	0.10	0.029%	0.014%	0.029%	69.9
1	450	460	10.00	5.0	0.10	0.033%	0.036%	0.052%	38.3
1	600	489	10.00	5.0	0.10	0.040%	0.072%	0.092%	21.6
1	750	513	10.00	5.0	0.10	0.050%	0.130%	0.155%	12.9
1	900	534	9.80	3.2	0.06	0.067%	0.223%	0.256%	7.8
1	1000	546	9.70	2.5	0.05	0.080%	0.302%	0.342%	5.9
1	1500	598	9.50	1.6	0.03	0.100%	1.093%	1.143%	1.7
1	2000	637	9.40	1.3	0.03	0.200%	2.853%	2.953%	0.7
5	850	527	10.00	3.2	0.06	0.067%	0.188%	0.221%	9.2
5	1250	574	9.70	2.5	0.05	0.100%	0.598%	0.648%	3.1
7	1800	622	9.50	1.6	0.03	0.200%	1.963%	2.063%	1.0

NOTES TO TABLE 8

1. Assumed minimum pH and estimated alkalinity and EPM from pH.

2. Higher hydroxide alkalinities may be recommended to reduce risk of silica deposits.